EUROPEAN CONTRACT LAW AND THE DIGITAL SINGLE MARKET

EUROPEAN CONTRACT LAW AND THE DIGITAL SINGLE MARKET

The Implications of the Digital Revolution

Edited by
Alberto De Franceschi

Cambridge – Antwerp – Portland

Intersentia Ltd
Sheraton House | Castle Park
Cambridge | CB3 0AX | United Kingdom
Tel.: +44 1223 370 170 | Fax: +44 1223 370 169
Email: mail@intersentia.co.uk
www.intersentia.com | www.intersentia.co.uk

Distribution for the UK and Ireland:
NBN International
Airport Business Centre, 10 Thornbury Road
Plymouth, PL6 7 PP
United Kingdom
Tel.: +44 1752 202 301 | Fax: +44 1752 202 331
Email: orders@nbninternational.com

Distribution for Europe and all other countries:
Intersentia Publishing nv
Groenstraat 31
2640 Mortsel
Belgium
Tel.: +32 3 680 15 50 | Fax: +32 3 658 71 21
Email: mail@intersentia.be

Distribution for the USA and Canada:
International Specialized Book Services
920 NE 58th Ave. Suite 300
Portland, OR 97213
USA
Tel.: +1 800 944 6190 (toll free) | Fax: +1 503 280 8832
Email: info@isbs.com

European Contract Law and the Digital Single Market. The Implications of the Digital Revolution
© The editor and contributors 2016

The authors have asserted the right under the Copyright, Designs and Patents Act 1988, to be identified as authors of this work.

No part of this book may be reproduced, stored in a retrieval system, or transmitted, in any form, or by any means, without prior written permission from Intersentia, or as expressly permitted by law or under the terms agreed with the appropriate reprographic rights organisation. Enquiries concerning reproduction which may not be covered by the above should be addressed to Intersentia at the address above.

Cover image © Tomas Griger

ISBN 978-1-78068-422-2
D/2016/7849/109
NUR 822

British Library Cataloguing in Publication Data. A catalogue record for this book is available from the British Library.

PREFACE

The EU is committed to making the single market fit for the digital age. This far-reaching political strategy has many facets. One consists in providing the Digital Single Market with a suitable legal infrastructure. A comprehensive and well-structured body of rules is required to enhance the protection of consumers and data subjects, while ensuring businesses the legal certainty they need to invest in this field and support growth and innovation.

The essays collected in this book, based on a series of seminars hosted by the Department of Law of the University of Ferrara between March and May 2016, address the impact of digital technology on private law within the EU. The papers examine a variety of topics, including the EU approach to personal information (namely as a tradeable commodity and as the object of a fundamental right for the individuals concerned), the protection of consumers' and users' rights and the issues surrounding the emergence of the so called platform economy.

The analysis, while being concerned to a large extent with contract law issues, extends to data protection and copyright law. Private international law issues are equally considered.

As the editor of this collection, I wish to thank the authors for their enthusiastic participation in this project.

A special thank also goes to the Alexander von Humboldt Foundation, whose financial support has been crucial to the organisation of the seminars and the publication of this volume.

Alberto De Franceschi
Ferrara – Munich, 25 July 2016

CONTENTS

Preface v
List of Authors xiii

European Contract Law and the Digital Single Market: Current Issues and New Perspectives
    Alberto De Franceschi 1

1. Digital Technology and Contract Law. 1
2. The Impact of Digital Technology on Private Law Relationships 3
3. Data as a Tradeable Commodity and the New Instruments for their Protection 5
4. Legislative Instruments for a Digital Single Market 8
5. New Features of Standard Contracts in the Digital Market. 11
6. Online Platforms in the 'Sharing Economy'. 15
7. Concluding Remarks 17

PART I.
THE IMPACT OF DIGITAL TECHNOLOGY ON PRIVATE LAW RELATIONSHIPS. 19

Disruptive Technology – Disrupted Law? How the Digital Revolution Affects (Contract) Law
    Christian Twigg-Flesner. 21

1. Introduction. 21
2. Disruptive Technology. 22
3. Law, Technological Development and Disruptive Effects 23
4. Designing Legal Rules for Disrupted Law. 27
5. The Disruptive Effect of the Digital Revolution. 28
6. Disrupted Law? 31
7. The EU and the Digital Revolution 42
8. A Concluding Thought. 47

PART II.
DATA AS A TRADEABLE COMMODITY AND THE NEW INSTRUMENTS FOR THEIR PROTECTION 49

Data as a Tradeable Commodity
Herbert ZECH 51

1. Data as the Object of a Contract 53
2. Existing Exclusive Rights for Different Kinds of Data 62
3. Introduction of a Data Producer Right? 74
4. Conclusion 79

Jurisdiction regarding Claims for the Infringement of Privacy Rights under the General Data Protection Regulation
Pietro FRANZINA 81

1. Introductory Remarks 82
2. Jurisdiction and the Right to an Effective Judicial Remedy 85
3. Jurisdiction under the Brussels I *bis* Regulation 88
4. Jurisdiction over the Infringement of Privacy Rights Pursuant to Article 79(2) GDPR 96
5. The Coordination between Article 79(2) GDPR and the Brussels I *bis* Regulation 103

PART III.
THE LEGISLATIVE INSTRUMENTS FOR A DIGITAL SINGLE MARKET 109

A European Market for Digital Goods
Michael LEHMANN 111

1. Digital Goods 112
2. Portability 113
3. The Supply of Digital Content 115
4. Online Trade in Goods 125
5. Summary 126

Supply of Digital Content. A New Challenge for European Contract Law
Reiner SCHULZE 127

1. Introduction 127
2. Current Change in Contract Law 131

3.	Conformity.	134
4.	Conceptual Continuity and Innovation: Further Examples	139
5.	Conclusion	143

Reflections on Remedies for Lack of Conformity in Light of the Proposals of the EU Commission on Supply of Digital Content and Online and Other Distance Sales of Goods
Geraint HOWELLS 145

1.	Introduction.	145
2.	Online and Distance Contracts	147
3.	Proposal for Digital Content Directive	155
4.	Conclusions	160

The Proposal of the EU Commission for a Regulation on Ensuring the Cross-Border Portability of Online Content Services in the Internal Market
Karl-Nikolaus PEIFER 163

1.	Introduction.	163
2	Portability – Why Do We Have to Regulate It?	164
3.	Why and How Do We Regulate Portability?	165
4.	Supporters and Critics of the Draft Proposal	166
5.	The Core and Content of the Draft Proposal	167
6.	The Function of the Regulation within the Digital Agenda.	171
7.	Possible Effects.	171

The Law Applicable to Consumer Contracts in the Digital Single Market
Peter KINDLER 173

1.	The Substantive Law Background.	173
2.	The Law Applicable to Consumer Contracts: General Outline	175
3.	The Key Connecting Factor: Activities 'Directed' to the Consumer Country (Article 6(1)(b) Rome I Regulation)	176
4.	Merely Indicative Facts	182
5.	The Need for Specific Conflicts of Law Rules for International Consumer Contracts in the Digital Single Market	183

PART IV.
NEW FEATURES OF STANDARD CONTRACTS IN THE DIGITAL MARKET ... 187

Standard Terms and Transparency in Online Contracts
Rodrigo MOMBERG ... 189

1. Introduction ... 189
2. Digital Content and Wrap Contracts ... 191
3. The Invisibility of Wrap Contracts ... 193
4. The Enforceability of Wrap Contracts ... 195
5. Transparency in EU Law ... 198
6. Transparency and Wrap Contracts ... 202
7. Curing Invisibility: Sufficient Notice and Specific Consent ... 204
8. Conclusions: The Unavoidable Assessment of Substantive (Un)Fairness ... 206

Contracts Concluded by Electronic Means in Cross-Border Transactions. 'Click-Wrapping' and Choice-of-Court Agreements in online B2B Contracts
Martin GEBAUER ... 209

1. Introduction ... 209
2. Normative Background ... 211
3. The European Court of Justice and Choice-of-Court Agreements Concluded by Electronic Means ... 213
4. Jurisdictional Consequences of the Decision Given by the ECJ in Case C-322/14 ... 217

PART V.
ONLINE PLATFORMS IN THE 'SHARING ECONOMY' ... 221

Crowdsourcing Consumer Confidence. How to Regulate Online Rating and Review Systems in the Collaborative Economy
Christoph BUSCH ... 223

1. Introduction ... 223
2. More Reputation, Less Regulation? ... 225
3. Recent Regulatory Initiatives ... 229
4. Key Elements of a Regulatory Framework for Reputation Systems ... 232
5. Conclusion ... 242

Online Dispute Resolution Platform. Making European Contract Law More Effective
 Jorge Morais Carvalho and Joana Campos Carvalho............ 245

1. Introduction... 245
2. Overview of Alternative Dispute Resolution in European Contract Law 247
3. Online Dispute Resolution Platform 250
4. Critical Analysis of the Legal Regime and its Implementation 263

LIST OF AUTHORS

Christoph Busch
Professor of German and European Private and Business Law, Private International Law, European Legal Studies Institute, University of Osnabrück, Germany

Joana Campos Carvalho
PhD student in Private Law, NOVA Faculty of Law, Lisbon, Portugal; Member of Centro de I&D sobre Direito e Sociedad (CEDIS); and Research Fellow at Fundação para a Ciência e a Tecnologia

Alberto De Franceschi
Assistant Professor of Italian Private Law and European Contract Law, University of Ferrara, Italy; Research Fellow of the Alexander Humboldt Foundation at the University Ludwig Maximilian of Munich, Germany

Pietro Franzina
Associate Professor of International Law, University of Ferrara, Italy

Martin Gebauer
Professor of Private Law, Private International Law and Comparative Law, University of Tübingen, Germany

Geraint Howells
Dean and Chair Professor of Commercial Law, City University of Hong Kong

Peter Kindler
Professor of Private Law, Business Law, Private International Law and Comparative Law, University of Munich, Germany

Michael Lehmann
Affiliated Research Fellow in Intellectual Property and Competition Law, Max Planck Institute for Innovation and Competition, Munich, Germany

List of Authors

Rodrigo Momberg
Associate Professor, Faculty of Law, Catholic University of Valparaiso, Chile; Visiting Research Fellow, Institute of European and Comparative Law, University of Oxford, United Kingdom

Jorge Morais Carvalho
Professor of Private Law, NOVA Faculty of Law, Lisbon, Portugal; Researcher at Centro de I&D sobre Direito e Sociedade (CEDIS)

Karl-Nikolaus Peifer
Professor of Civil Law with Copyright Law, Industrial Property Protection, New Media and Economic Law, University of Cologne, Germany; Director of the Institute of Media and Communications Law, University of Cologne, Germany

Reiner Schulze
Professor of German and European Civil Law, University of Münster, Germany

Christian Twigg-Flesner
Professor of Commercial Law, University of Hull, United Kingdom

Herbert Zech
Professor of Life Sciences Law and Intellectual Property Law, University of Basel, Switzerland

EUROPEAN CONTRACT LAW AND THE DIGITAL SINGLE MARKET

Current Issues and New Perspectives

Alberto De Franceschi

1. Digital Technology and Contract Law 1
2. The Impact of Digital Technology on Private Law Relationships......... 3
3. Data as a Tradeable Commodity and the New Instruments for their Protection 5
4. Legislative Instruments for a Digital Single Market................... 8
5. New Features of Standard Contracts in the Digital Market 11
6. Online Platforms in the 'Sharing Economy' 15
7. Concluding Remarks.. 17

1. DIGITAL TECHNOLOGY AND CONTRACT LAW

Digital technology has significantly changed the balance in society and economic relationships, offering new opportunities for innovative business models. This raises challenging questions affecting several aspects of the law.[1] Scholars, practitioners, policy-makers and legislators are therefore involved in an ongoing debate and need to find appropriate answers in order to react adequately to the challenges of the digital revolution. This book aims to chart, analyse and clarify some of the main questions and issues.

For a functioning market economy, private law has to provide a general framework and efficient tools. The realisation of a connected Digital Single Market is one of the ten priorities of the European Commission, which aims to react appropriately to the challenges of the digital revolution in order to use this opportunity for economic growth. In the framework of its Digital Single

---

[1] See R. Schulze and D. Staudenmayer, 'Digital Revolution – Challenges for Contract Law' in R. Schulze and D. Staudenmayer (eds.), *Digital Revolution: Challenges for Contract Law in Practice*, Nomos – Hart, Baden-Baden – Oxford, 2016, p. 19; C. Twigg-Flesner, in this volume, p. 21 et seq; N. Helberger, M.B.M. Loos, L. Guibault, C. Mak and L. Pessers, 'Digital Content Contracts for Consumers' (2013) 36 *Journal of Consumer Policy (JCP)* 37 et seq.

Market Strategy, the European Commission announced a set of measures aiming to create better access to digital goods and services across Europe for both consumers and businesses, underlining that the absence of consistent EU-wide criteria creates barriers to entrance, hinders competition and reduces predictability for investors throughout Europe.[2]

The harmonisation of European private law is a challenge of increasing significance[3] and has inspired important developments, as shown for example by the European Commission's release, on 9 December 2015, of three legislative proposals as a part of its 'Digital Single Market Strategy': the Proposal for a Directive on certain aspects concerning contracts for the supply of digital content,[4] the Proposal for a Directive on certain aspects concerning contracts for the online and other distance sales of goods,[5] and the Proposal for a Regulation on ensuring the cross-border portability of online content services in the internal market.[6] Since then, on 27 April 2016, EU Regulation 2016/679 on the protection of natural persons with regard to the processing of personal data and on the free movement of such data has been adopted.[7]

Furthermore, at the end of May and at the beginning of June 2016, the European Commission launched another wave of communications[8] and proposals (the Proposal for a 'geo-blocking' Regulation, aimed at preventing

---

[2] 'A Digital Single Market Strategy for Europe', COM(2015) 192 final.

[3] See D. WALLIS, 'Digital agenda: the role of the legal community in helping the EU legislator' (2016) 5 *Journal of European Consumer and Market Law (EuCML)* 1 et seq.

[4] COM(2015) 634 final. See C. TWIGG-FLESNER, in this volume, p. 43 et seq.; H. ZECH, in this volume, p. 55; M. LEHMANN, in this volume, p. 115 et seq.; R. SCHULZE, in this volume, p. 131 et seq.; G. HOWELLS, in this volume, p. 155 et seq.; P. KINDLER, in this volume, p. 174 et seq. Cf. e.g. G. SPINDLER, 'Verträge über digitale Inhalte' (2016) 19 *Multimedia und Recht (MMR)*, p. 147, 219; M. WENDLAND, 'GEK 2.0? Ein Europäischer Rechtsrahmen für den Digitalen Binnenmarkt' (2016) 13 *Zeitschrift für das Privatrecht der Europäischen Union (GPR)* 8 et seq. See also *infra*, n. 5.

[5] COM(2015) 635 final. See C. TWIGG-FLESNER, in this volume, p. 43 et seq.; M. LEHMANN, in this volume, p. 125 et seq.; R. SCHULZE, in this volume, p. 128 et seq.; G. HOWELLS, in this volume, p. 149 et seq.; P. KINDLER, in this volume, p. 174 et seq. Cf. on both directive proposals e.g. J.M. SMITS, 'New European Proposals for Distance Sales and Digital Contents Contracts: Fit for Purpose?' (2016) 24 *Zeitschrift für Europäisches Privatrecht (ZEuP)*, p. 319 et seq.; M. SCHMIDT-KESSEL, K. ERLER, A. GRIMM and M. KRAMME, 'Die Richtlinienvorschläge der Kommission zu Digitalen Inhalten und Online-Handel – Teil I' (2016) 13 *GPR*, p. 2 et seq.; M. SCHMIDT-KESSEL, K. ERLER, A. GRIMM and M. KRAMME, 'Die Richtlinienvorschläge der Kommission zu Digitalen Inhalten und Online-Handel – Teil II' (2016) 13 *GPR* 54 et seq. See also the studies for the the JURY-Committee of the European Parliament by H. BEALE, B. FAUVARQUE-COSSON, V. MAK and J.M. SMITS, in www.europarl.europa.eu.

[6] COM(2015) 627 final. See K.-N. PEIFER, in this volume, p. 163 et seq.; M. LEHMANN, in this volume, p. 111 et seq.

[7] [2016] OJ L119. See in this regard H. ZECH, in this volume, p. 66 et seq.; P. FRANZINA, in this volume, p. 81 et seq.; C. TWIGG-FLESNER, in this volume, p. 41.

[8] See e.g. the Communication from the Commission of 25.5.2016 'Online Platforms and the Digital Single Market – Opportunities and Challenges for Europe', COM(2016) 288/2 final and the Communication from the Commission of 2.6.2016 'A European agenda for the collaborative economy', COM(2016) 320 final. See C. BUSCH, in this volume, p. 223 et seq.

discrimination against customers by traders on the grounds of the customer's residence when accessing websites or ordering goods or services[9] and a Proposal for a Regulation dealing with cross-border parcel delivery services[10]), as well as a number of consultations on, for example, 'the regulatory environment for platforms, online intermediaries, data and cloud computing and the collaborative economy'.[11]

It is therefore of crucial importance to assess how the digital revolution is affecting the development of the law and to provide appropriate answers to the resulting new challenges, including in particular the impact of digital technology on private law relationships, the status of information as a tradeable commodity and the new instruments to protect it, the legislative instruments for a Digital Single Market concerning consumers' and users' rights when buying goods, services or digital content online, the new features of standard contracts in the digital market, and the platform economy and online dispute resolution platforms.

2. THE IMPACT OF DIGITAL TECHNOLOGY ON PRIVATE LAW RELATIONSHIPS

Technological developments are having an increasingly significant impact on private law relationships. New business opportunities and smart technologies are altering the equilibrium of the established paradigms that form the basis of many legal rules. In particular, several questions have arisen as a result of those changes, going beyond the national level to also affect European and international issues. As a consequence of the impact of the digital revolution on the legal system and of the broad scope of the Digital Single Market Strategy for Europe, the analysis in this book has not been limited to 'pure' contract law issues, but also includes such topics as data protection law, copyright law, public regulation of the infrastructure of the modern information society and international private law.

An accurate analysis of the impact of the digital revolution on contract law, as well as of the positive and negative aspects of introducing new rules in response to new developments, is therefore crucial.[12] Many problems can be solved by shaping the interpretation and application of the existing European and national rules according to the new needs of the 'digital world'. There are in any case situations in which it may be necessary to reshape existing laws to a

---

[9] COM(2016) 289 final.
[10] COM(2016) 285 final.
[11] See C. TWIGG-FLESNER, in this volume, p. 35; see also C. BUSCH, in this volume, p. 225 et seq.
[12] See C. TWIGG-FLESNER, in this volume, p. 23 et seq. See also European Commission, 'Guidance on the Implementation/Application of Directive 2005/29/EC on Unfair Commercial Practices', SWD(2016) 163 final.

greater or lesser extent or to simplify certain issues, or furthermore to adopt new rules for new situations which are not yet regulated by the existing rules.[13]

Nevertheless, there comes a point that developments are so ground-breaking that it is no longer possible to adapt existing legal rules to new circumstances and it is therefore necessary to reshape them. In doing so, there is the real risk that in order to respond quickly to the new developments the legislator will draw up new rules without adequately weighing all the relevant elements of the new market relationships, and therefore that, in the rush to provide a legal response, more problems might be created than solved.[14] In addition, rules that are too stringent and limiting may have a disruptive effect on technological progress and on economic developments.[15]

In this context, a number of questions and challenges have been raised regarding possible legal responses.[16] In particular, the regulation of digital content, which can either be acquired as a stand-alone transaction or be supplied as a part of the acquisition of a physical item, has to be dealt with. It is therefore first of all necessary to categorise digital content[17] and to draft a general framework to deal appropriately with situations where digital content is not of the required quality or causes damage to other digital content, physical devices, or data belonging to the user.[18]

Furthermore, the question needs to be addressed as to whether the already existing legislative instruments are appropriate in order to cope with the needs of the digital economy, in which in many cases the supplier of goods and services is not a 'trader' according to the common understanding of this concept. In this volume are therefore discussed the criteria for distinguishing between different categories of market players in the digital environment.[19] In relation to this, it has been underlined that many consumers have nowadays become 'prosumers',[20] offering not only goods but also services such as transportation, accommodation

---

[13] SCHULZE and STAUDENMAYER, 'Digital Revolution – Challenges for Contract Law', *supra* n. 1, p. 31; see also WALLIS, *supra* n. 3, p. 1 et seq.
[14] See C. TWIGG-FLESNER, in this volume, p. 22.
[15] See 'Commission Communication on Online Platforms and the Digital Single Market – Opportunities and Challenges for Europe', COM(2016) 288/2, p. 5, which stresses that: 'The collaborative economy is a good example where rules designed with traditional and often local service provision in mind may impede online platform business models'. See R. PODSZUN and S. KREIFELS, 'Digital platforms and competition law' (2016) 5 *EuCML*, 33 et seq.
[16] See C. TWIGG-FLESNER, in this volume, p. 27 et seq.
[17] Cf. C. LANGHANKE and M. SCHMIDT-KESSEL, 'Consumer Data as Consideration' (2015) 4 *EuCML* 218; S. VAN ERP, 'Ownership of digital assets?' (2016) 5 *EuCML* 73.
[18] See C. TWIGG-FLESNER, in this volume, p. 31.
[19] See C. TWIGG-FLESNER, in this volume, p. 37, who underlines that 'the long-held paradigm transaction of the one-shot trader consumer contracts is on increasingly shaky ground'.
[20] The concept of 'prosumer' is due to C. BUSCH, H. SCHULTE-NÖLKE, A. WIEWIÓROWSKA-DOMAGALSKA and F. ZOLL, 'The Rise of the Platform Economy: A New Challenge for EU Consumer Law?' (2015) *EuCML* 4.

or cleaning services. This often happens for example in the context of contracting via online platforms.

A further result of the digital revolution is the 'Internet of things'[21], which is linked to the concept of objects that can communicate with one another. A typical example of this concept in operation is the use of robotics in smart factories[22], household apps, or self-driving cars. This field offers huge development potential, but several issues, especially in relation to liability, need to be clarified.[23]

3. DATA AS A TRADEABLE COMMODITY AND THE NEW INSTRUMENTS FOR THEIR PROTECTION

In the data-driven economy, data is handled both as a tradeable commodity[24] and as a counter-performance[25]: transactions concerning data involve both contract and property law. This raises new issues and requires tailor-made answers.

Information plays a key role in the dynamics of the digital revolution. From an economic perspective it is therefore relevant to know whether and what kind of data content can be protected. The fundamental question is therefore whether data can be recognised in law as 'protectable rights'. In the digital world, data are in fact an important '*res intra commercium*', namely tradeable goods, the legal protection of which even today remains the subject of considerable debate. More recently, the problem of deleting data on the Internet and the 'right to be forgotten' has been discussed in connection with search engines and social networks.[26] Such discussion informs also the background of Regulation 2016/679 on the protection of natural persons with regard to the processing of personal

---

21 N. HELBERGER, 'Profiling and Targeting Consumers in the Internet of Things – A New Challenge for Consumer Protection' in SCHULZE and STAUDENMAYER, *supra* n. 1, p. 135 et seq.
22 E. PALMERINI and A. BERTOLINI, 'Liability and Risk Management in Robotics' in SCHULZE and STAUDENMAYER, *supra* n. 1, p. 189.
23 European Commission, 'A Digital Single Market Strategy for Europe', COM(2015) 192 final. See on this point C. TWIGG-FLESNER, in this volume, p. 30 et seq.; R.H. WEBER, 'Contractual Duties and Allocation of Liability in Automated Digital Contracts' in SCHULZE and STAUDENMAYER, *supra* n. 1, p. 163; C. WENDEHORST, 'Consumer Contracts and the Internet of Things' in SCHULZE and STAUDENMAYER, *supra* n. 1, p. 189; H. ZECH, 'Gefährdungshaftung und neue Technologien' (2013) 68 *Juristenzeitung (JZ)* 21; P. BRÄUTIGAM and T. KLINDT, 'Industrie 4.0., das Internet der Dinge und das Recht' (2015) 68 *Neue Juristische Wochenschrift (NJW)* 1137.
24 Cf. A. DE FRANCESCHI and M. LEHMANN, 'Data as Tradeable Commodity and the new Measures for their Protection' (2015) 1 *The Italian Law Journal* 51.
25 LANGHANKE and SCHMIDT-KESSEL, *supra* n. 17, p. 218 ff.
26 See most recently K. KOWALIK-BAŃCZYK and O. POLLICINO, 'Migration of European Judicial Ideas Concerning Jurisdiction over Google on Withdrawal of Information' (2016) 17 *German Law Journal*, p. 315 et seq.

data and on the free movement of such data.[27] The most relevant applications in this context are databanks (e.g. Big Data)[28] and computer files, especially those that not only have the ability merely to organise data but that also to modify them.

Directive 2011/83/EU on consumer rights[29] already protects data as such, gives special treatment to the commercial use of data in sales law, and expressly protects 'digital content', which is defined in Article 2(11) (definitions) as 'data which are produced and supplied in digital form'. Pursuant to Articles 9 and 16(m) of the Directive, consumers may also be precluded from exercising their usual withdrawal right where digital content is supplied online pursuant to long-distance and off-premises contracts.[30]

In the later withdrawn proposal for a Common European Sales Law (CESL)[31], digital content was defined in Article 2(j) as 'data which are produced and supplied in digital form, whether or not according to the buyer's specifications, including video, audio, picture or written digital content, digital games, software and digital content which makes it possible to personalise existing hardware or software'. It is significant that, pursuant to Article 5(b) CESL, digital data were put on fundamentally the same footing as any other object that might be purchased, irrespective of whether they are delivered online or offline or made available for download. The result is that, in cross-border EU commercial transactions, digital data are treated as tradeable goods and are legally protectable in the same manner as other goods. This also applies to laws relating to interference in the performance of obligations, as illustrated by Article 106 et seqq. CESL concerning the buyer's remedies.

In its ground-breaking decision *UsedSoft v. Oracle,*[32] the Court of Justice of the European Union also came to a similar conclusion, although the decision of course only directly addressed the problem of exhaustion with respect to the sale of software. However, in line with the principle of a European Single Market, the

---

[27] Regulation 2016/679 of the European Parliament and of the Council of 27 April 2016 on the protection of natural persons with regard to the processing of personal data and on the free movement of such data, and repealing Directive 95/46/EC (General Data Protection Regulation) [2016] OJ L119.

[28] See e.g. V. MAYER-SCHÖNBERGER and C. CUKIER, *Big Data: A Revolution That Will Transform How We Live, Work and Think,* Houghton Mifflin Harcourt, New York 2013, *passim*; J. WOLF, 'Der rechtliche Nebel der deutsch-amerikanischen "NSA-Abhöraffäre"' (2013) 68 JZ 1039.

[29] Directive 2011/83/EU of 25 October 2011 on consumer rights [2011] OJ L304/64.

[30] See also recital 19 of Directive 2011/83/EU concerning the withdrawal right. Cf. R. SCHULZE and J. MORGAN, 'The Right of Withdrawal' in G. DANNEMANN and S. VOGENAUER (eds.), *The Common European Sales Law in Context,* Oxford University Press, Oxford 2013, p. 294.

[31] COM(2011) 635 final.

[32] Case C-128/11, *UsedSoft GmbH v. Oracle International Corp.*, judgment of 3 July 2012; see H. ZECH, 'Vom Buch zur Cloud' (2013) 5 *Zeitschrift für Geistiges Eigentum* 368; A. OHLY, 'Anmerkung zu EuGH v. 3.7.2012, Rs. C-128/11, Usedsoft/Oracle' (2013) 68 JZ 42 et seqq.; R. HILTY, K. KÖKLÜ and F. HAFENBRÄDL, 'Software Agreements: Stocktaking and Outlook – Lessons from the UsedSoft v. Oracle Case from a Comparative Law Perspective' (2013) 44 *International Review of Intellectual Property and Competition Law (IIC)* 263.

Court fundamentally treated data that a user ultimately transfers on an outright basis as tradeable goods to be dealt with in a manner akin to property under commercial law.[33] Although this argument was initially derived by the Court from Directive 2009/24/EU on the legal protection of computer programs[34], it must also be applied to other digital content (data in electronic form) that is handled and sold in EU cross-border transactions like goods in respect of which 'ownership'[35] can be transferred.

The foregoing also applies in relation to downloads from the 'cloud', in other words when in the course of cloud computing data are sold on an outright basis to a buyer to both use and own.[36] In practice, a data set in this context is commercially treated as a good and should therefore be classified and treated as such under commercial law principles.

Therefore, the Consumer Rights Directive, the withdrawn CESL and the Court of Justice of the European Union have all already clearly delivered an important contribution to the commercial handling of data in a digital form. When data usage rights are the subject of an outright sale, there is a transfer of both sale or gift objects and rights. From a contractual and property law perspective, the provisions relating to the sale or gifting of objects apply. These principles can as a consequence also apply to the transfer of intellectual and industrial property in digital form.

Article 3 of the Commission's Proposal for a Directive on certain aspects concerning contracts for the supply of digital content[37], concerning the Directive's scope of application, provides for the possibility that in contracts for the supply of digital content the consumer may actively provide counter-performance other than money, in the form of personal data or other data.[38] Facing the innovative character of this provision it is first of all necessary to clarify the factual and legal situation underlying contracts concerning data. In this volume, Herbert Zech provides therefore a deeper analysis of the concept

---

[33] See point 61 of the judgment *UsedSoft GmbH v. Oracle International Corp.*: 'It should be added that, from an economic point of view, the sale of a computer program on CD-ROM or DVD and the sale of a program by downloading from the internet are similar. The on-line transmission method is the functional equivalent of the supply of a material medium'; similarly M. LEHMANN, in U. LOEWENHEIM (ed.), *Handbuch des Urheberrechts*, 2nd ed., C.H. Beck, Munich 2010, p. 1866, (12).
[34] Directive 2009/24/EC of 23 April 2009 on the legal protection of computer programs, [2009] OJ L111/16.
[35] See point 46 of the judgment *UsedSoft GmbH v. Oracle International Corp.*
[36] Cf. M. LEHMANN, 'Immaterialgüterrechtliche Aspekte' in G. MEENTS and J.G. BORGES (eds.), *Cloud Computing*, C.H. Beck, Munich 2016, p. 456.
[37] COM(2015) 634 final.
[38] See C. TWIGG-FLESNER, in this volume, p. 40; H. ZECH, in this volume, p. 59 et seq.; R. SCHULZE, in this volume, p. 141; VAN ERP, *supra* n. 17, p. 73; LANGHANKE and SCHMIDT-KESSEL, *supra* n. 17, p. 218 et seq.; SCHMIDT-KESSEL, ERLER, GRIMM and KRAMME, *supra* n. 5, p. 55, 57; WENDEHORST, *supra* n. 23, p. 193; WENDLAND, *supra* n. 4, p. 8, 12. Cf. most recently F. FAUST, 'Digitale Wirtschaft – Analoges Recht: Braucht das BGB ein Update?' (2016) 70 *Neue Juristische Wochenschrift – Beilage (NJW-Beil)* 29 et seq.

of data as a commodity and the interplay between factual exclusivity and legal exclusivity, the legal framework for contracts concerning different kinds of data, and the arguments for and against the introduction of a data producer right.[39]

The issues surrounding privacy rights and the law of data protection are further examined by Pietro Franzina from a private international law perspective. Specifically, the latter's contribution is concerned with the rules laid down in Regulation (EU) 2016/679 (the General Data Protection Regulation) to identify the court, or courts, that possess jurisdiction over actions against data controllers and processors relating to the alleged infringement of the rights enshrined in the Regulation, and on the coordination of those rules with the general EU rules regarding jurisdiction in civil and commercial matters.[40] While the subject-matter of the above proceedings is not contractual in nature, the rules on data protection are likely to exert a significant influence on the behaviour of the various actors of the Digital Single Market and on the contractual practice they are involved in.

4. LEGISLATIVE INSTRUMENTS FOR A DIGITAL SINGLE MARKET

The digital revolution requires the existing legal rules to be adapted to deal with new circumstances. So long as these rules have a sufficient degree of flexibility, this is relatively straightforward.[41] However, for many of the new issues arising from the latest forms of contracting, such adaptation is not possible. As a result, a flood of new instruments and proposals have been published since December 2015 in the context of the Digital Single Market Strategy.

In particular, while the Proposal for a Directive on the supply of digital content has been widely welcomed[42], the Proposal for a Directive on the online and other distance sales of goods has been sharply criticised, due to its overlap with the Directive 2011/83/EU on consumer rights and Directive 99/44/EC on consumer sales.[43] As underlined in these two proposals[44], the withdrawn CESL[45], instead of now becoming superfluous, plays a key role in the background

---

[39] See in particular H. ZECH, in this volume, p. 51 et seq.
[40] See P. FRANZINA, in this volume, p. 81 et seq.
[41] C. TWIGG-FLESNER, 'Innovation and EU Consumer Law' (2005) 28 *Journal of Consumer Policy* 409.
[42] COM(2015) 635 final. See C. TWIGG-FLESNER, in this volume, p. 42 et seq.; M. LEHMANN, in this volume, p. 115, 126; R. SCHULZE, in this volume, p. 127, 143; G. HOWELLS, in this volume, p. 145, 161; SCHULZE and STAUDENMAYER, 'Digital Revolution – Challenges for Contract Law', *supra* n. 1, p. 31 et seq.
[43] In this sense, see e.g. G. HOWELLS, in this volume, p. 160.
[44] See e.g. COM(2015) 634 final, p. 2; COM(2015) 635 final, p. 2.
[45] COM(2011) 635 final. See in this regard e.g. R. SCHULZE (ed.), *Common European Sales Law – Commentary*, Beck – Nomos – Hart, Munich – Baden-Baden – Oxford 2012; G. DANNEMANN

and in the current reality of European contract law, as it already contains some important answers to the challenges of the digital age even if it did not survive to the legislative process.[46] Many of the provisions contained in the two proposals are indeed rooted in the draft CESL (although the digital content proposal in particular does contain some significant modifications).[47]

After Michael Lehmann provides an assessment of the key issues in the aforementioned proposals[48], the focus moves to the supply of digital content as well as to the associated challenges for several core aspects of contract law.[49] In this respect, Reiner Schulze looks into the development of contractual doctrine that is necessary in accompanying the legislation.[50] For instance, the development of the Internet has already led to lively discussions on whether traditional contract principles and concepts need to be redefined or could continue to be applied to new circumstances in e-commerce.[51] Although the continued application of existing principles and concepts may in some instances have been apt ten years ago, 'digitalisation' has brought about developments such as 'automated' conclusion of contract in the 'Internet of things', and therefore has given rise new circumstances for which traditional principles and concepts are no longer apt.[52] The analysis thus focuses on whether European contract law doctrine needs further development in relation to the supply of digital content. The author discusses specific points such as conformity as a central concept of European contract law (as well as its extension to include 'interoperability' and 'accessibility') and the context of this new understanding of conformity in other aspects of contract law, such as remedies and restitution after termination.[53] At a wider scale, the supply of digital content also impacts on the notion of contract and types of contract in European law, for example by shifting the emphasis from sales to the single exchange of performance, thus affecting the *pacta sunt servanda* principle, and by bringing long-term contracts further to

---

and S. VOGENAUER, *supra* n. 30; O. REMIEN, S. HERRLER and P. LIMMER (eds.), *Gemeinsames Europäisches Kaufrecht für die EU?*, Beck, Munich, 2012; H. SCHULTE-NÖLKE, F. ZOLL, N. JANSSEN and R. SCHULZE (eds.), *Der Entwurf für ein optionales europäisches Kaufrecht*, Sellier, Munich 2012; M. SCHMIDT-KESSEL (ed.), *Ein einheitliches europäisches Kaufrecht?*, Sellier, Munich 2012.

[46] R. SCHULZE, 'The New Shape of European Contract Law' (2015) 4 *EuCML*, p. 139, 141.

[47] Relating to the significance of the withdrawn CESL for the further development of EU law, see C. TWIGG-FLESNER, in this volume, p. 44; R. SCHULZE, in this volume, p. 128, 143. See also R. SCHULZE and F. ZOLL, *European Contract Law*, Beck – Nomos – Hart, Munich – Baden-Baden – Oxford 2016, p. 286.

[48] M. LEHMANN, in this volume, p. 111 et seq.

[49] R. SCHULZE, in this volume, p. 127 et seq.

[50] See R. SCHULZE, in this volume, p. 131.

[51] See e.g. R. BRADGATE and G. HOWELLS, 'When Surfers Start to Shop: Internet Commerce and Contract Law' (1999) 19 Legal Studies 287, 314.

[52] See also C. TWIGG-FLESNER, in this volume, p. 30.

[53] On the concept of conformity in the new directive proposals see also SCHMIDT-KESSEL, ERLER, GRIMM and KRAMME, 'Die Richtlinienvorschläge der Kommission zu Digitalen Inhalten und Online-Handel – Teil II', *supra* n. 5, p. 65.

the fore. In light of these new features, Schulze assesses the way in which the technological and economic shift caused by the digital age marks a new phase in the development of European contract law.[54]

In light of the proposals of the EU Commission on supply of digital content and on online and other distance sales of goods, remedies for lack of conformity are thus subject of a deep and multifaceted analysis by Geraint Howells.[55] Remedies have indeed been the most controversial aspect of the EU engagement with sales law. In particular the move to maximum harmonisation has brought to the fore the tension between regimes that favour giving the seller the opportunity to cure and those that reserve to the consumer the right to reject non-conforming goods.[56] The European Commission is now, for a third time, trying to establish a mandatory regime with a hierarchy of remedies in the proposed Directive relating to online and other distance sale of goods. This choice has been strongly criticised as being insensitive to the concerns of consumers and the desire for respect of local traditions. By contrast, it seems to be more reasonable for the EU to address the issue of contracts for the supply of digital content, as the relevant legal regime is unclear in many Member States and underlines that in this context a right to cure may be more legitimate. Furthermore, it is worth drawing a contrast between the EU proposal and the recent rules established for digital content in the UK's Consumer Rights Act 2015.[57]

A key role in the Digital Single Market Strategy is furthermore played by the proposal for a Regulation on ensuring the cross-border portability of online content services in the internal market.[58] It is thus necessary to examine the regulatory challenges in this area. Focusing on contract law as well as on intellectual property rights, Karl-Nikolaus Peifer underlines that the aforementioned proposal is a limited but necessary step forward, which will help to build a unified digital market by increasing consumer trust and fostering new business models which are more suited to the digital world.[59]

As a consequence of the far-ranging impact of the digital revolution on the legal system and of the broad scope of the Digital Single Market Strategy for Europe, the analysis in this volume has not been limited to 'pure' contract law issues, but also includes such topics as international private law with a specific focus on the law applicable to consumer contracts in the digital single

---

[54] R. SCHULZE, in this volume, p. 131 et seq. Cf. also C. TWIGG-FLESNER, in this volume, p. 42 et seq.
[55] See G. HOWELLS, in this volume, p. 145 et seq.
[56] These circumstances led to the abandonment of the reform of sales law in the Consumer Rights Directive and to the withdrawal of the Common European Sales Law.
[57] See G. HOWELLS, in this volume, p. 150. Cf. C. TWIGG-FLESNER, in this volume, p. 33 and R. SCHULZE, in this volume, p. 128.
[58] COM(2015) 627 final.
[59] See K.-N. PEIFER, in this volume, p. 163 et seq.

market. In both the proposed Directive for the supply of digital content and the proposed Directive on certain aspects concerning contracts for the online and other distance sale of goods, the European Commission has underlined that the provisions in these Directives should not prejudice the application of the rules of private international law, in particular of Regulation 593/2008 (Rome I) and Regulation 1215/2012 (Brussels I *bis*).[60] The Commission therefore seems convinced that the existing private international law instruments are already fit for the digital world and that they respond adequately to the needs of Internet transactions. In view of the mentioned legislative self-restraint, Peter Kindler assesses the usefulness and meaningfulness of the Commission's approach in this context.[61]

5. NEW FEATURES OF STANDARD CONTRACTS IN THE DIGITAL MARKET

In the digital environment, as well as in physical marketplaces, buyers and users are faced with standard-form contracts for the use or acquisition of goods and services. Often the non-drafting party assents to terms they are not aware of. As in particular concerns contracts concluded by electronic means, the user is frequently not even aware he is entering a legal or contractual relationship at all or that the legal relationship will entail an obligation to pay. The fact that standard terms in online contracts are not easily visible and/or legible for the user creates new challenges for the adequate protection of non-drafting parties, for instance because of a different understanding of the principle of transparency and of the assessment of the user's the intention to be bound by the contract or by some of its terms.[62] It is thus necessary to examine whether the current and proposed EU legislation (in particular in the context of the Digital Single Market Strategy) is adequate to cope with these challenges.[63]

Looking at Directive 2011/83/EU on consumer rights (CRD), it is in particular possible to put forward some thoughts about the European legislator's methodological approach to standard-form contracts concluded by electronic means. This assessment could be useful in order to understand which criticisms EU legislator should avoid in taking the next steps within its Digital Single Market.

---

[60] See e.g. recital 49 of the proposal for a Directive on certain aspects concerning contracts for the supply of digital content.
[61] See P. KINDLER, in this volume, p. 173 et seq.
[62] Cf. H.W. MICKLITZ, 'Chapter 3. Unfair Terms in Consumer contracts' in N. REICH, H.-W. MICKLITZ, P. ROTT and K. TONNER, *European Consumer Law*, 2nd ed., Intersentia, Cambridge 2014, p. 142, who has described transparency as the formal element that must be assessed with regard to the unfairness of standard terms.
[63] See R. MOMBERG, in this volume, p. 189 et seq.

Given the large number of complaints about 'Internet cost traps'[64], specific 'formal requirements' for contracts concluded by electronic means were already included in the CRD in order to enable consumers to be clearly and succinctly informed about all the costs before entering into a binding contract. Article 8(2) CRD sets out some specific formal requirements for contracts concluded by electronic means which entail an obligation to pay. It prescribes that if the trader does not comply with the 'formal requirements' provided for contracts stipulated by electronic means, the consumer 'shall not be bound by the contract or the order'. The solution, however, has not received unanimous praise. The 'button solution' aimed to provide stronger and more effective protection for the consumer than that ensured by the right of withdrawal, by introducing a unitary solution to unfair behaviour by the trader in relation to 'Internet cost traps'.

First, the distinction between the 'different' formal requirements contained in Article 8 CRD – 'clear and legible', 'easily legible' and 'clear and prominent' – is not that precise.[65] This can of course cause problems in assessing how and whether, in the concrete situation, the trader complied with the requirements.

Second, the formulation contained in Article 8(2) CRD that 'the consumer is not bound by the contract or order' is rather vague. It could be inferred that, since the consumer is not bound by the order, no contract has been concluded. In particular, it is not clear whether, if the trader does not comply with the 'formal requirements', he nevertheless has to fulfil his contractual obligations and whether the consumer may keep the goods he receives without having to pay a reasonable price for them. Furthermore, it is not clear whether, in the event of non-fulfilment of the 'formal requirements', the provision on inertia selling (Article 27 CRD) could be applied, as neither the contract nor the order is binding on the consumer.[66]

The second point of criticism has wider ramifications. The concrete meaning of the notion of 'non-bindingness' is unclear in the formulation of Article 8 CRD. In addition, the difference between the 'non-bindingness' of the whole contract

---

[64] Council of the European Union, 4 June 2010, Internet cost traps – Directive on Consumer Rights – Information from the German delegation <http://register.consilium.europa.eu/pdf/en/10/st10/st10604.en10.pdf>, p. 3: 'Given the constantly large number of complaints about *Internet* cost traps … Germany has proposed the inclusion of a "button solution" in the Directive on Consumer Rights … With a view to efficient consumer protection in Europe, Germany asks the Member States and the Commission to support the proposal on a 'button solution', thereby increasing consumer confidence in the Internal Market'.

[65] See in this concern the critique by G. HOWELLS and J. WATSON, 'Article 25' in SCHULZE (ed.), *supra* n. 45, p. 169: 'The variation in formal requirements serves as an ideal example of the failure to adopt overarching standards. The approach in the DCFR of summarising the form requirements in one article that is applicable to all pre-contractual information duties is preferable. This creates legal certainty for the traders who have to fulfil the pre-contractual obligations'.

[66] A. DE FRANCESCHI, 'Informationspflichten und formale Anforderungen im Europäischen E-Commerce' (2013) 62 *GRUR Int.* 865.

according to Article 8(2) CRD and the 'non-bindingness' of the term according to Article 25(2) CRD is rather vague.[67] The implementation of the CRD in the EU Member States as a result presents a very fragmented picture.[68]

Concerning contracts stipulated by electronic means that do not comply with the aforementioned 'formal requirements', some national legislators merely reproduced the neutral wording of the CRD, providing that in such cases the consumer 'shall not be bound by the contract or order'. This is what happened in Austria, Belgium, Estonia, Ireland, Italy, Spain, Sweden and the United Kingdom. Furthermore, if the trader has not complied with the 'formal requirements', according to Dutch law the contract is voidable, whereas it is null and void in Finland, France and Luxembourg. The German and Polish legislators have expressly provided that in such cases the contract has to be considered not to have been concluded. It is doubtful whether this choice can be considered consistent with the CRD as it does not take into account the situation in which the consumer wants to adhere to the contract. In particular, some national implementing rules appear not to be consistent with the CRD. The provisions laid down by the Austrian and German legislators exclude, for example, from the scope of application of the implementing provisions of Article 8(2) CRD contracts concluded exclusively by means of individual communication (e.g. email or SMS). Some other EU Member States have provided also that in the event of non-fulfilment of the duty to make the consumer aware, directly before he places his order, in a clear and prominent manner, of the information provided in Article 6(1)(a), (e), (q) and (r) CRD, the consumer 'shall not be bound' by the contract. This solution is clearly not consistent with the CRD, which merely provides that the contract or the order is not binding in the event of the violation of the 'formal requirements' provided in Article 8(2) CRD.

This shows that it is not really efficient for the system to prohibit Member States from maintaining or introducing in their national laws 'provisions diverging from those laid down in a full harmonisation Directive, including more or less stringent provisions to ensure a different level of consumer protection', unless setting clear and definite parameters.[69]

Particularly in the digital market, uncertainty concerning private law consequences is neither admissible nor sustainable: the speed of transactions is

---

[67] Art. 25(2) CRD: 'Any contractual terms which directly or indirectly waive or restrict the rights resulting from this Directive shall not be binding on the consumer'.

[68] See e.g. M. LOOS, 'Implementation of CRD (Almost) Completed, Harmonisation Achieved?' (2014) 3 *Zeitschrift für Europäisches Unternehmens- und Verbraucherrecht / Journal of European Consumer and Market Law (euvr)*, p. 213; G. DE CRISTOFARO and A. DE FRANCESCHI, 'Consumer Sales in Europe', Intersentia, Cambridge, 2016.

[69] This consistency between the use of standard terminology and achievement of a real harmonisation is indeed of crucial relevance – especially in the web environment – in pursuing the goal in building a solid infrastructure for a functioning digital single market.

decisive and is partly dependent on legal certainty. It remains crucial, therefore, to eliminate legal obstacles to cross-border transactions. In the light of this, an analysis of the problems that have prevented true full harmonisation on the basis of the CRD and other EU Directives is a useful starting point to ensure that such criticisms are not repeated in the drafting of new instruments in the framework of the Digital Single Market Strategy for Europe.[70]

We can thus learn an important lesson from the problems arising from the implementation of CRD: that there is a need for EU institutions to set out not only harmonised sanctions but also harmonised sanction regimes. Otherwise, the EU legislator merely sets general definitions, without being able to achieve real harmonisation. If harmonised sanction regimes are not established in combination with harmonised sanctions, the only allegedly 'harmonised definition' – namely that of harmonised sanctions – will be filtered through a lens coloured by the categories of sanctions in the Member States' national legal systems. This undoubtedly leads more to fragmentation than to harmonisation, as can easily be seen from the various solutions adopted in the Member States in implementing the CRD.

The European legislator may therefore now take the opportunity to create – at least for the legislative instruments proposed in the framework of the Digital Single Market Strategy, but hopefully also in the future, relating to a broader scope of application – a system of clear and consistent sanctions and to explain in detail how they work, in order to start building a coherent system of the grounds for invalidity in EU contract law. Otherwise, the European Court of Justice will (still be constricted to) 'dress in the clothes of the legislator',[71] a situation which almost inevitably leads to the familiar fragmented and contradictory results.

The Digital Single Market Strategy also contains measures aiming to create better access to digital goods and services across Europe for both consumers and businesses, underlining that the absence of consistent EU-wide criteria creates entrance barriers, hinders competition and reduces predictability for investors throughout Europe.[72] In the same document the European Commission underlined that, at present, markets are largely domestic in terms of online services, and that only a small number of small and medium-

---

[70] See 'A Digital Single Market Strategy for Europe', COM(2015) 192 final, p. 10: 'Delivering the conditions for a true single market by tackling regulatory fragmentation to allow economies of scale for efficient network operators and service providers and effective protection of consumers' is one of the main goals actually pursued by the EU legislator. Cf. H. SCHULTE-NÖLKE, 'The Brave New World of EU Consumer Law – Without Consumers, or Even Without Law?' (2015) 4 *EuCML* 135, 137.

[71] The expression is due to O. POLLICINO, 'The Sense of the Court of Justice of the European Union for Digital Privacy: Interpretation or Manipulation?', presented at the conference 'Building the Legal Framework of the Digital Single Market', Brussels (European Parliament, 2 July 2015).

[72] 'A Digital Single Market Strategy for Europe', COM(2015) 192 final.

sized European businesses sell cross-border. In this volume, after a chapter by Rodrigo Momberg dealing with standard terms and transparency in online contracts, Martin Gebauer analyses therefore the topic of formal requirements in cross-border contracts concluded online, dealing in particular with issues of choice-of-law and jurisdiction.[73] As European Union law only addresses the formal requirements of forum selection clauses, the analysis focuses specifically on the formal requirements of these procedural contracts and on choice-of court-agreements, also discussing the content and the consequences of the European Court of Justice's first decision, given in 2015, on the formal requirements of a choice-of-court agreement concluded by electronic means.

6. ONLINE PLATFORMS IN THE 'SHARING ECONOMY'

The rise of the 'platform economy' is having a disruptive effect on both established business models and the associated legal rules.[74] In recent years, the case law and legislative activity concerning electronic platforms have taken off considerably in most European Member States. Important problems and questions continue to arise not only concerning private law issues, but also regarding competition and public law, as well as several aspects of administrative law and tax regulation.

Following the communication of the European Commission on online platforms[75], we can now expect even increasing debate and legislative activity. Indeed, the existing law does not seem to be able to provide adequate solutions to several further problems relating to the so-called 'platform economy'. It remains in particular unclear whether or not the platform should be considered an intermediary between those who offer and those who require goods or services.

---

[73] M. GEBAUER, in this volume, p. 209 et seq.
[74] See C. BUSCH, H. SCHULTE-NÖLKE, A. WIEWIÓROWSKA-DOMAGALSKA and F. ZOLL, 'The Rise of the Platform Economy: A New Challenge for EU Consumer Law?' (2016) 5 *EuCML* 3 et seq. Cf. also the further papers collected in the issue 1/2016 of *EuCML*, which is entirely dedicated to the topic of the online platforms and collects the results of the conference 'Platform Services in the Digital Single Market' held in Osnabrück on 19–20 November 2015 and organised by C. Busch, H. Schulte-Nölke, A. Wiewiórowska-Domagalska and F. Zoll. See also M. COLANGELO and V. ZENO-ZENCOVICH, 'Online Platforms, Competition Rules and Consumer Protection in the Travel Industry' (2016) 5 *EuCML*, p. 75 et seq.; D.S. EVANS and R. SCHMALENSEE, *Matchmakers: The New Economics of Multisided Platforms*, Harvard Business Review Press, Boston 2016.
[75] See in particular 'Commission Communication on Online Platforms and the Digital Single Market – Opportunities and Challenges for Europe', COM(2016) 288/2, p. 5, which stresses that: 'The collaborative economy is a good example where rules designed with traditional and often local service provision in mind may impede online platform business models'.

The contradictory recent legislative and judicial developments at national and European level[76] highlight the need for more tailor-made rules, especially relating to the multilateral relationships that characterise platform services. The current lack of clear answers at European level shows that a homogeneous and systematic approach is required in order to avoid fragmented solutions at national level. Issues like security of buyers of good or services, privacy, discrimination in the choice of clients, employment conditions[77] and payment are indeed common at least in all EU Member States and therefore need a coherent approach by the EU legislator. In order to achieve this goal, an appropriate cost-benefit analysis is crucial, especially from the perspective of introducing a unitary legislative framework at European level and of finding an adequate balance between conflicting interests like freedom of establishment, free provision of services, competition and contract law issues, thereby ensuring better protection for both consumers and businesses.

Relating to this fundamental topic, this volume contains an analysis of the need to adjust EU consumer contract law to take into account the changing market structure caused by the rise of online platforms. In particular, Christoph Busch assesses how reputation systems can contribute to the development of generalised trust among users of sharing economy platforms and thus complement existing regulatory tools for enhancing consumer confidence.[78]

Online platforms also play a significant role as regards the enforcement of contractual rights.[79] The existence of simple, efficient, fast and low-cost ways of resolving disputes is key to making European contract law more effective. With this goal in mind, on 21 May 2013 a legislative package was adopted including Directive 2013/11/EU on alternative dispute resolution for consumer disputes (Directive on Consumer ADR) and Regulation 524/2013 on online dispute resolution for consumer disputes (Regulation on Consumer ODR). Jorge Morais Carvalho and Joana Campos Carvalho analyse therefore the ODR platform created by Regulation 524/2013, in force since 15 February 2016 in most Member States. Their chapter starts with an overview of dispute resolution and then carries out a legal analysis of Regulation 524/2013, emphasising the main legal issues arising from the ODR platform and

---

[76]   Concerning the issues related to the Uber platform, see the request for a preliminary ruling from the Juzgado Mercantil No 3 de Barcelona (Spain) lodged on 7 August 2015 – *Asociación Profesional Élite Taxi v. Uber Systems Spain, S.* (Case C-434/15). For the full text of the decision of the referring court, see Juzgado Mercantil No 3, 17 June 2015, Asunto 929/2014D2, available at: <www.poderjudicial.es>. Cf. the request for a preliminary ruling from the Rechtbank van Koophandel Brussel (Belgium) lodged on 5 October 2015 – *Uber Belgium BVBA v. Taxi Radio Bruxellois NV* (Case C-526/15).

[77]   See e.g. V. DE STEFANO, 'The rise of the "just-in-time workforce": On-demand work, crowdwork and labour protection in the "gig-economy"' <www.ilo.org>.

[78]   See C. BUSCH, in this volume, p. 223 et seq.

[79]   See J. MORAIS CARVALHO and J. CAMPOS CARVALHO, in this volume, p. 245 et seq.

concluding with a critical evaluation of the legal regime and its practical implementation.[80]

7. CONCLUDING REMARKS

The digital revolution requires the existing legal rules to deal with new circumstances. So long as these rules have a sufficient degree of flexibility, this is relatively straightforward.[81] However, for many of the new issues arising from the latest forms of contracting, an adaptation by way of interpretation is not possible. As a result, a flood of new instruments and proposals has been published in the framework of the Digital Single Market Strategy. The EU is indeed committed to making the single market fit for the digital age. This far-reaching political strategy has many facets. One consists in providing the Digital Single Market with a suitable legal infrastructure. A comprehensive and well-structured body of rules is required to enhance the protection of consumers and data subjects, while ensuring businesses have the legal certainty they need to invest in this field and support growth and innovation.

Most of the questions raised in relation to the digital revolution concern aspects which cannot be left merely to national scholars, judges or legislators, but rather which need to be addressed in a coherent way at the European level and with an interdisciplinary approach. The main goal of the European legal community is to contribute to the analysis and development of European private law – multifaceted as it is – in a way that takes into consideration the impact of the digital revolution. It is indeed of crucial importance to find the right balance, one that does not leave new markets without rules, but one that also avoids overregulation in order both to ensure consumer and user protection as well as to let new business models flourish. This could set a fundamental milestone for the development of a true European single market: by taking a coherent approach, the EU legislator can contribute to strengthening transparency and legal certainty, thereby enhancing the attractiveness of internal and cross-border digital commerce for all everyone involved, including actors from outside the EU.

---

80   J. Morais Carvalho and J. Campos Carvalho, in this volume, p. 263.
81   Twigg-Flesner, *supra* n. 42.

PART I
THE IMPACT OF DIGITAL TECHNOLOGY ON PRIVATE LAW RELATIONSHIPS

DISRUPTIVE TECHNOLOGY – DISRUPTED LAW?

How the Digital Revolution Affects (Contract) Law

Christian TWIGG-FLESNER

1. Introduction 21
2. Disruptive Technology 22
3. Law, Technological Development and Disruptive Effects 23
4. Designing Legal Rules for Disrupted Law 27
5. The Disruptive Effect of the Digital Revolution 28
6. Disrupted Law? 31
   6.1. Regulation of Digital Content 31
   6.2. The Consumer Notion – When Does a Consumer Become a Trader? 34
   6.3. Platforms and Intermediaries 36
   6.4. Combination of Sale, Service and Digital Content in One Transaction but Different Parties 37
   6.5. The Internet of Things and Automated Contracting 38
   6.6. The Use of Personal Data 40
   6.7. Other Questions 42
7. The EU and the Digital Revolution 42
   7.1. Proposals on Contract Law Aspects 43
   7.2. Other Proposals 45
   7.3. What Next? 46
8. A Concluding Thought 47

1. INTRODUCTION

Much has been said about the effect of disruptive technology on business. In this contribution, an attempt is made to consider, in general terms, the implications of 'disruptive technology' for the law, particularly contract law.[1] The particular disruptive technology focused on in this contribution is more of a 'disruptive development': the so-called digital revolution, and the new business opportunities and production methods which have emerged from the increasing digitalisation of so many activities, not least by utilising the potential of the Internet combined with smart-technology. These developments undoubtedly pose interesting challenges for contract law, particularly established paradigms forming the basis of many legal rules. This analysis begins by exploring the notion of disruptive technology, before considering the general challenges for, and possible responses by, the law as a result of new developments in technology or business practice. It will then highlight the main novelties of the digital revolution and turn to some of the specific legal issues which the digital revolution seems to create and consider potential legal responses.

The key argument of this contribution is that there is a danger of rushing towards introducing new legal rules in response to new developments without rigorous consideration of the specific issues for both businesses and consumers which are created by things such as the digital revolution. Once these issues have been fully scoped, any legal responses need to be calibrated carefully so as to deal with these issues in a focused manner – there is a risk that, in the rush to provide a legal response, more problems might be created than solved.

2. DISRUPTIVE TECHNOLOGY

Before considering the impact of the digital revolution on contract law, a few words should be said about the meaning of 'disruptive technology', a term used frequently in this context. This notion has gained prominence in the writings of Clayton Christensen,[2] and focuses on the way technological developments can affect the way existing business models operate. In brief, Christensen distinguishes between two types of technological evolution: first, there is 'sustaining technology', by which he means technology which is evolving gradually or simply improving established technologies, particularly their performance. In contrast, 'disruptive technology' is a new type of technology, which, when first introduced, might be less reliable than established technologies, but will become reliable rapidly. Its possible applications are uncertain when it first appears, but once it has gained a certain critical mass of recognition, it can significantly affect the way things are done. At that point, the disruptive effect of such new technology materialises, and established ways of conducting

---

[1] Many of the other contributions to this collection will provide a more detailed discussion of specific issues.
[2] Seminally, C.M. CHRISTENSEN, *The Innovator's Dilemma – When New Technology Causes Great Firms to fail*, Harvard Business Review Press, Boston 1997.

business are threatened by new business models which take advantage of the new technology. In Katyal's words:

> 'Disruptive innovation goes beyond improving existing products; it seeks to tap unforeseen markets, create products to solve problems consumers don't know they have, and ultimately to change the face of the industry.'[3]

Without doubt, the creation of the Internet and the rise of mobile smart-technology have had a disruptive effect. Take e-commerce as an example: initially, there were cautious attempts by businesses to utilise the Internet as an alternative trading mechanism to physical stores, but soon, new businesses emerged online to compete with established bricks-and-mortar traders. Over time, the disruptive effect has become apparent, as many famous brands have disappeared altogether or moved from a significant physical presence to an online presence only.[4]

Since then, the appeal of the Internet and digital technology has broadened, and ever-new opportunities for developing new business models utilising digital technology are emerging. This raises interesting questions for various branches of the law: be it the protection of intellectual property rights in digital content, the protection of personal data, or consumer rights when buying goods, services or digital content online. Many of these issues are discussed by legal scholars and policy-makers alike, and there is an ongoing debate about what sort of legal response to the digital revolution might be required. This chapter is an attempt to chart some of the main questions and issues which will need to be addressed by lawmakers.

3. LAW, TECHNOLOGICAL DEVELOPMENT AND DISRUPTIVE EFFECTS

It is a trite observation that technology is continuously evolving – indeed, continuous innovation is supported by government policies at all levels.[5] The key implications of this for the design of legal rules are generally recognised: lawmakers need to ensure that legal rules are adaptable to new circumstances, whether that be new products or new ways of doing business. This can be

---

[3] N. KATYAL, 'Disruptive Technologies and the Law' (2014) 102 *Georgetown Law Journal* 1685.
[4] In the United Kingdom, several famous names have disappeared from the High Street, including Woolworths, Comet, Habitat and MFI. See SKY NEWS, *What's Disappeared From The UK High Street?*, 25 April 2016 <http://news.sky.com/story/1684696/whats-disappeared-from-the-uk-high-street> accessed 19.05.2016.
[5] For example, the United Kingdom government has established a dedicated agency to support innovation (Innovate UK – see <www.innovate.gov.uk>), and the EU runs its Innovation Union initiative <http://ec.europa.eu/research/innovation-union/index_en.cfm>.

achieved by putting into place legal rules which have an inherent degree of flexibility and can therefore be deployed in circumstances which might not have been foreseen at the time when these rules were adopted.[6] One example of this is the EU's Unfair Commercial Practices Directive (2005/29/EC), which combines outlawing some established unfair commercial practices (those listed in the Annex to the Directive) with flexible general prohibitions which set general criteria against which the fairness of commercial practices can be assessed.[7] The use of such open-textured prohibitions provides a high degree of 'future-proofing'.

However, whilst a degree of future-proofing in designing legal rules is possible, there will come a point when developments have got to the stage where existing legal rules reach the limits of their adaptability to new circumstances, and at that stage, it will become necessary to consider changes to existing legal rules. At this juncture, disruptive technology therefore not only disrupts established business models, but also existing legal rules. As Copps puts it,

> 'Technology, it's almost trite to say, is developing at a blistering pace and forces us to confront new issues – to think anew and to act anew. How do the legal and regulatory frameworks apply and keep pace? How do regulators make good decisions in this fast-moving, paradigm-shifting environment?'[8]

The pace of technological development creates a serious risk of an overly hasty legal response. However, even in the face of rapidly changing circumstances, care needs to be taken in working out what an appropriate legal response might be. There is a danger that the fear of a legal vacuum will prompt a rush towards putting into place legal rules which might seem to tackle new issues created by technological development but which soon turn out to be unsuitable or entirely unworkable. At the same time, there is the opposing risk that waiting for too long could create regulatory gaps which could result in unacceptable levels of consumer detriment.

So the challenge for the law and regulation to keep up with technological developments is a serious one. Bennet Moses summarises her careful analysis of this issue thus:

> 'Our metaphors of law struggling to keep pace with technology reflect an important truth: as technology changes, legal dilemmas arise. As technological change becomes increasingly rapid, the need for a methodical response to these problems becomes

---

[6] C. TWIGG-FLESNER, 'Innovation and EU Consumer Law' (2005) 28 *Journal of Consumer Policy* 409.
[7] See e.g. H.W. MICKLITZ, 'The General Clause on Unfair Practices' in G. HOWELLS, H.W. MICKLITZ and T. WILHELMSSON, *European Fair Trading Law*, Ashgate, Aldershot 2006.
[8] M.J. COPPS, 'Disruptive Technology... Disruptive Regulation' [2005] *Michigan State Law Review* 309.

increasingly urgent. We need to closely analyse the roles played by different legal institutions and the methodologies they adopt in easing the law's transition to the future.'[9]

However, in all of this, it is important not to lose sight of the fact that not only might new rules be required to deal with new issues which arise from technological development – one also needs to be aware that existing rules and regulatory procedures could impede the adoption of technological advances by businesses and consumers alike,[10] and so sometimes the challenge is not only to make sure that legal rules appropriate to new technological developments are in place, but also that established rules and procedures are revised or removed so as not to stifle the exploitation of the potential offered by technological developments. For example, some time ago, Copps identified as one challenge the need to create a legal 'landscape which really fosters innovation'[11] whilst not losing sight of how consumers can benefit. This might require some courage in developing the law into new territory rather 'than spending all of our time trying to shoe-horn new technologies into old regulatory categories'.[12] In other words, once technological developments have become disruptive, innovation in legal rules needs to follow. That entails a recognition that legal rules which might have served business and consumers well in the past, even in the light of incremental technological advances, cannot be unduly strained to cover issues resulting from disruptive technological developments.

In this regard, Koopman *et al.*[13] have suggested that instead of extending existing regulatory regimes to new business models, it could be considered whether existing rules should be relaxed to put incumbents and new entrants on a deregulated level playing field. However, such an approach would entail the serious risk that existing protective rules, especially for consumers, are eroded and should not be the starting point; nevertheless, when it comes to assessing existing legal rules, it is worthwhile considering whether these rules still fulfil their intended purpose.

The disruptive effect of a particular development can be gauged by considering whether a specific issue can be dealt with by applying existing legal rules to the particular questions which have been identified in respect of that issue. Thus, if it is possible to maintain existing rules but to clarify how these should be applied in the context of a new development, then the disruptive effect to the law is minimal – indeed, such an approach might reflect the robust

---

[9] L. BENNETT MOSES, 'Recurring Dilemmas: The Law's Race to Keep up with Technological Change' [2007] *University of Illinois Journal of Law, Technology and Policy* 239, 285.
[10] Cf. C. KOOPMAN, M. MITCHELL and A. THIERER, 'The Sharing Economy and Consumer Protection Regulation – The Case for Policy Change' (2014–15) 8 *Journal of Business Entrepreneurship and Law* 529, 530.
[11] COPPS, *supra* n. 8, p. 309.
[12] *Ibid.*, p. 312.
[13] KOOPMAN, MITCHELL and THIERER, *supra* n. 10.

design of existing legal rules and their potential for extending these to new circumstances.[14]

However, this may not always be sufficient to tackle new issues adequately. The next option would therefore be to consider whether minor reforms such as clarifications or modifications of key definitions, modifications to the wording of existing legal rules and extension of provisions determining the scope of existing legal rules would be sufficient to deal with a specific new issue. A principle commonly invoked in this context is 'functional equivalence', i.e. to identify the key features of the developments governed by existing rules and to set out how these would be transferred into the context of any new development.[15] As long as such minor changes are enough, there is still no real disruptive effect on the law – even if there might already be a strong disruptive effect on business.

But if this still does not manage to ensure that the necessary and appropriate legal rules are in place, then it will be necessary to consider more far-reaching reforms. At this point, the law itself will start to feel a disruptive effect. Thus, once it has become clear that one cannot simply tweak existing rules to make them fit new developments, thought will have to be given to how new rules should be designed. This requires a careful evaluation of the nature of this new development and the particular legal issues it raises, before determining what sort of approach would provide suitable legal rules. It may be possible to develop new legal rules by analogy with existing legal rules – as discussed below, new rules for the supply of digital content might follow comparable rules applicable to the sale of goods but be modified and supplemented as is necessary to reflect the distinct nature of this type of transaction. This would take the law into new territory but could still be regarded as a progressive development, although in some legal systems, such an approach might start to feel disruptive.

However, if even such an approach is not possible and a completely new set of legal rules is required, perhaps reflecting new principles or doctrines, then there is a truly disruptive effect on the law. At this point, the earlier warning about not rushing towards adopting legal rules, even in the face of rapid developments in technology and/or business models, needs to be heeded. It is in the very nature of disruptive developments that things will happen at a rapid pace but issues might be felt acutely at one point, only to disappear and be replaced by other issues which have a more lasting effect. The latter types of issue need to be the focus for new legislation, once there is some clarity with regard to the specific

---

[14] The European Commission updated its guidance document on the application of the Unfair Commercial Practices Directive (2005/29/EU) to explain how the existing rules operate in the digital environment, e.g. with regard to online platforms. See European Commission, *Guidance on the Implementation/Application of Directive 2005/29/EU on Unfair Commercial Practices*, SWD(2016) 163 final.

[15] For example, the United Nations Convention on the Use of Electronic Communications in International Contracts 2005 deploys this principle in clarifying how rules in existing conventions such as the Convention on the International Sale of Goods 1980 (CISG) would operate in the electronic environment.

matters which require legal rules and time has been taken to analyse what the appropriate legal responses should be. This is the point at which swift proposals for the introduction of new legal rules should be made.

This is subject to the caveat that a rush towards new regulation might not only result in rules which turn out to be unsuitable, but it could also have the detrimental effect of stifling innovation which utilises disruptive technology to develop new business models or applications. So a trade-off is needed between avoiding the law lagging too far behind and allowing disruptive effects to settle before updating the law.

4. DESIGNING LEGAL RULES FOR DISRUPTED LAW

In the previous section, it was argued that a disruptive technological development such as the digital revolution also has a disruptive effect on the law once it is no longer possible to apply existing legal rules to new circumstances, or to modify the scope of existing legal rules to ensure that these address new issues brought about by technological changes. Once that point has arrived, it will be necessary to identify what the particular features are that require new legal rules, and then to determine how to draft new legal rules which will deal with the particular issues associated with these features. However, if the disruptive effect is particularly severe, then it will be necessary to do more than merely draft new legal rules dealing with a specific problem that has been identified. It may also be necessary, even essential, to consider whether the introduction of new legal rules also requires a reconsideration of fundamental principles underpinning existing law. For instance, contract law is underpinned by the notion of freedom of contract and the exercise of private autonomy, and the many detailed doctrines and principles reflect this notion, whether by enhancing it or imposing limitations to it. However, as will be seen below, one consequence of the digital revolution is the rise of automated contracting where an individual interacts with a computer, or even two computers interacting with each other, to arrange a transaction. Does the fact that there is no mutual exercise of private autonomy mean that no contract has been concluded? Is it necessary to change the basic principle for treating a binding obligation as a contract to accommodate automation? Or should such situation not be treated as involving contracts at all, with new principles and rules having to be developed to provide a rational and coherent legal framework for this situation? This is, of course, a simple example, and few people will seriously suggest that this is anything but a contract,[16] but

---

[16] Cf. R. WEBER, 'Contractual Duties and Allocation of Liability in Automated Digital Contracts' in R. SCHULZE and D. STAUDENMAYER (eds.), *Digital Revolution – Challenges for Contract Law* Nomos, Baden-Baden 2016.

it does nevertheless illustrate the point that once has law been disrupted as a result of technological developments, it is important to consider how deeply that disruptive effect is felt.

When it comes to developing concrete legal rules to respond to a new development, it will also need to be considered whether these should be focused narrowly to address the specific issue, or whether it is possible to develop legal provisions with a broader scope which could extend to as yet unforeseen circumstances. This is a familiar question – is it preferable to respond with specific and precise rules, or is it better to lay down broad standards which require concretisation when applied to particular circumstances?[17] This will depend on factors such as whether the issues which have been identified are sufficiently contained to be amenable to clear rules, whether standards should be set to guide businesses in utilising disruptive technology in the future, whether there is identified consumer detriment that needs to be addressed, and so on. There are other matters, but the ones mentioned here are enough to underline the importance of being thorough and prompt, but not rushed, when developing a legal response to the disruptive effects felt by technological developments.

The ideas from this and the previous section need to be explored in the context of the particular challenges of the digital revolution, so in the following section the disruptive effects of the digital revolution and the way in which these might also disrupt current law are explored.

5. THE DISRUPTIVE EFFECT OF THE DIGITAL REVOLUTION

This section outlines three key areas in which the digital revolution is likely to have a disruptive effect both on established business models and the associated legal rules. As noted previously, the disruptive element of the digital revolution is brought about by a combination of the wide availability of fast Internet connections, the popularity of smartphone technology and the consequent advances in both digital content and hardware. These developments make it possible to utilise the digital environment and digital content to do things which previously had to be done by different means.

Initially, the digital revolution largely opened up the possibility to use the Internet for online trading to complement the way of shopping for goods in physical stores, soon followed by individuals utilising the Internet to offer goods or services to others, and digital mechanisms ('online platforms') to facilitate this have evolved. In its basic form, an online platform is an intermediary bringing together recipients and suppliers of goods, services and

---

[17] For a classic discussion of this topic, see L. KAPLOW, 'Rules versus Standards: an Economic Analysis' (1992) 42 *Duke Law Journal* 557.

digital content, thereby making it easier for both to conclude transactions. Some platforms provide additional services such as payment processing, as well transaction guarantees and dispute resolution procedures in case something goes wrong. Some go further and set down detailed rules according to which products can be offered by a supplier on such a platform. Generally speaking, online platforms are a key means for suppliers who do not wish to maintain their own website to reach more customers, as well as enhancing the number of online channels for existing businesses. These platforms are often referred to as 'marketplace' platforms.

The rise of online platforms has prompted the development of a new type of business model which allows individuals or businesses to 'share' some of their assets with others. Thus, individuals can now let out their spare bedroom to visitors or arrange car-shares. There are many opportunities for bringing together people (whether private individuals or businesses) who have an underused asset with others who might wish to use that asset, and mobile technology has made it much easier to do this. This phenomenon – sharing underused assets with others – is often referred to as the 'sharing economy', or 'peer-to-peer' (P2P) economy (when suppliers and recipients are both non-traders). Although there has been considerable interest in ride-sharing and accommodation-sharing platforms, which primarily operate in a P2P or B2C setting, the sharing approach has also been utilised in the context of manufacturing (with a 'manufacturer' using production facilities operated by other businesses) and logistics/transport. Online platforms are at the heart of this development, and there is an increasing focus on the role of such 'sharing platforms' or 'collaborative economy platforms'.[18]

A second development of the digital revolution is the increased technical sophistication of many appliances which were previously predominantly mechanical or based on basic electronic technology. Thus, most household devices are equipped with digital technology which operates the mechanical elements (whether a refrigerator, washing machine or central heating system). In addition, many such devices can now be connected to the Internet through WiFi technology, which has made it possible to control multiple devices from one control station (such as a smartphone). In turn, these devices can collect data about performance and share this data with the user, the manufacturer or a third party. Also, in a commercial setting, the increasing automation of warehousing facilities combined with the Internet allows devices to identify when stocks are running low and to order new stock automatically. The possibility for devices to share data online and for this data to be processed and utilised is referred to

---

[18] See e.g. KOOPMAN, MITCHELL and THIERER, *supra* n. 10; S.R. MILLER, 'First Principles for Regulating the Sharing Economy' (2016) 53 *Harvard Journal on Legislation* 147, and European Commission, *Communication on a European Agenda for the Collaborative Economy*, COM (2016) 356 final.

as the 'Internet of things'.[19] It seems that the spread of the 'Internet of things' is progressing at a slower pace than anticipated,[20] but this might be beneficial from a legal-regulatory perspective, because there will be less urgency to react. Whether the potential of the Internet of things will ever be realised on a broad scale therefore remains somewhat uncertain, not least because costs seem to be too high, and interoperability of connected devices is still limited due to the lack of a standardised communications protocol.[21]

A third significant development is the combination of a mechanical process with digital design technology to create a new form of manufacturing technology: additive layer manufacturing, more popularly known as 3D printing.[22] Using this technology, it is now possible to manufacture goods by consecutively 'printing' very thin layers of an item to manufacture a finished physical item. There are a number of different ways in which additive layer manufacturing works, but there are common features: the design for the item to be 3D printed is supplied in a computer file created by using appropriate computer-aided design (CAD) software. The file is then sent to a physical device, the 3D printer, which manufactures a physical item based on the CAD file. This technology offers a number of new practical applications: it allows for customisation of products to the requirements of each buyer, items can be made to order (especially when there would not be sufficient demand for a large production run), and items can be ordered from anywhere in the world but created as close as possible to the location of the buyer (including the buyer's own 3D printer). Moreover, design files can be traded separately from the physical item: some websites allow a designer (whether a professional or a private individual) to create a design and upload this, thereby allowing others to buy a copy of the design or to order a printed version of the design from that website. Although this development has a huge potential for changing the way goods are made and customised to an individual buyer's preference, it also raises a host of legal issues. Literature on the legal implications of 3D printing has initially focused on the intellectual property concerns,[23] but this was soon followed by discussion of the product

---

[19] For a useful exploration of the potential of the Internet of things, see S. GREENGARD, *The Internet of Things*, MIT Press, Cambridge (MA) 2015.
[20] See 'Where the smart is', *The Economist* (11 June 2016), pp. 65–66.
[21] *Ibid.*
[22] See generally, C. ANDERSON, *Makers – The New Industrial Revolution*, Random House, London, 2012, and C. BARNATT, *3D Printing – The Next Industrial Revolution*, ExplainingTheFuture, Marston Gate 2013.
[23] See e.g. D.R. DESAI and G.N. MAGLIOCCA, 'Patents, Meet Napster: 3D Printing and the Digitisation of Things' (2014) 102 *Georgetown Law Journal*, 1691; A. LEWIS, 'The Legality of 3D Printing: How Technology is Moving Faster than the Law' (2014) 17 *Tulane Journal of Technology and Intellectual Property* 303; and P. REDDY, 'The Legal Dimension of 3D Printing: Analyzing Secondary Liability in Additive Layer Manufacturing' (2014) 16 *Columbia Science and Technology Law Review* 222.

liability implications if 3D-printed goods cause harm,[24] and wider debates about the private law implications.[25]

This brief account of three major developments brought about by the digital revolution illustrates the disruptive effect on existing business models and paradigm contracts. In turn, these developments also create questions as to what sort of legal response is needed to ensure that they are recognised in law and also that legal rules such as those on consumer protection developed in the context of contracts for the supply of goods and services between a trader and a consumer are not sidestepped by such new business models. The following section will consider a number of areas where existing legal rules may need to be revisited, or where a new legal response might be needed.

6. DISRUPTED LAW?

The various developments outlined above have given rise to a number of recurring questions about possible legal responses. This section will set out a number of particular legal issues which have been raised and consider possible solutions. As suggested in section 3 above, some aspects might seem less disruptive than others, and their degree of disruptiveness might vary between different legal systems.

6.1. REGULATION OF DIGITAL CONTENT

At the heart of the digital revolution is the way digital content is utilised: the range of applications ('apps') for smartphones is vast, digital content controls the way many physical devices operate, and CAD files provide designs which can be turned into physical items through 3D printing technology. Digital content can be acquired as a stand-alone transaction, for example by downloading an app onto a mobile device, purchasing software (such as an office package) for a computer, or buying design files. In addition, digital content may be supplied as part of the acquisition of a physical item (e.g. software controlling functions of a car).

With digital content having assumed such a central function, inevitable questions arise as to whether there are legal rules in place which deal with the

---

[24] N.D. BERKOWITZ, 'Strict Liability for Individuals? The impact of 3-D printing on Products Liability Law' (2014) 92 *Washington University Law Review* 1019.
[25] L.S. OSBORN, 'Regulating Three-Dimensional Printing: The Converging World of Bits and Atoms' (2014) 51 *San Diego Law Review* 553; C. TWIGG-FLESNER, 'Conformity of 3D prints – Can current Sales Law cope?', and G. HOWELLS and C. WILLETT, '3D Printing: The Limits of Contract and Challenges for Tort', both in R. SCHULZE and D. STAUDENMAYER (eds.), *Digital Revolution – Challenges for Contract Law*, Nomos, Baden-Baden 2016.

quality and fitness for purpose of such content, remedies when digital content is not of sufficient quality, and even for the consequences of digital content causing damage to other digital content, physical devices, or data stored belonging to the user of the digital content.

This is, of course, not an altogether new question: the classification of computer software has been a much-debated issue, and the current questions about the status of digital content (which includes 'old-fashioned' computer software) are not particularly new. However, the prevalence of digital content now has given new urgency to this question. Hitherto, most legal systems have debated whether the supply of software/digital content should be treated as a contract for the supply of goods, or one for the supply of services. Neither seems to make particularly good sense: digital content is by its nature intangible, and whilst it has in the past often been supplied on a physical medium, is increasingly transferred as a download via the Internet. Moreover, a transaction involving the supply of digital content does not involve the transfer of ownership over a physical asset (as would be the case with a contract for the sale of goods) but merely the right to acquire a copy of the digital content (unless the content is accessed online without being transferred to the user's device) with a licence to use this in accordance with the terms of the licence.[26]

In any case, it seems that these debates about how to classify digital content were only necessary for two reasons: (i) some legal systems operate with a closed category of contracts types; and (ii) more significantly, out of a desire to ensure that a person acquiring digital content could be assured that there was a quality standard required by law with appropriate remedies in circumstances where the digital content fails to reach that standard.[27]

In jurisdictions where the first reason applies, law reform may therefore be required to create a new type of contract for the supply of digital content. This may be more easily done in some jurisdictions than in others, and may also depend on whether this would be limited to certain categories of contracts (e.g. consumer contracts only) or be generally applicable. There are good reasons for having a separate category of contracts for the supply of digital content: as already mentioned, the supply of digital content has features which are distinct from other types of contract, and might even be said to combine elements of a range of different contracts (e.g. a service element, a hire element if the licence to use the software is time-limited, etc.).

---

[26] Cf. S. ARNERSTÅL, 'Licensing digital content in a sale of goods context' (2015) 10 *Journal of Intellectual Property Law and Practice* 750.

[27] Although there are English cases grappling with this issue, it has never been a major difficulty, because the common law has always been more flexible in recognising new *sui generis* contracts for the supply of goods which also include terms regarding the quality and fitness for purpose of the goods akin to those in the Sale of Goods Act 1979. See e.g. the recent Supreme Court ruling in *PST Energy 7 Shipping LLC v OW Bunker Malta Ltd* [2016] UKSC 23, para. [31].

With regard to the second reason, in particular, the question therefore arises whether specific legal rules for digital content should be introduced into law to set standards as to the quality and fitness for purpose of digital content, combined with appropriate remedies for circumstances where the digital content falls below that standard or where the interaction of the digital content with physical devices or other digital content causes loss or damage. If it is accepted that this is should be the objective, then there are a number of things to consider: first, the relevant provisions could be aligned as closely as possible with the corresponding rules applicable to the sale of goods. A somewhat crude way of achieving this would be simply to extend the definition of 'goods' in national law to include 'digital content', as was done for example in New Zealand.[28] The obvious drawback of doing this is that rules designed for tangible items might not be entirely suitable for intangible items.

So perhaps a better option might be to use the relevant rules on quality, fitness for purpose and remedies applicable to goods as a starting point, but modify them as might seem appropriate for digital content. This was the approach adopted in the UK's Consumer Rights Act 2015 (CRA). In the CRA, the requirements applicable to goods have been borrowed in almost unchanged form to introduce requirements corresponding to these in contracts for the supply of digital content. Thus, digital content has to be of satisfactory quality,[29] fit for a particular purpose,[30] and be as described.[31] Similar, although not fully identical, remedies are also provided: initially, a consumer can choose between repair and replacement of the digital content, with a second-stage remedy of price reduction (up to the full value of the price paid).[32] Damages for additional losses can also be recovered.[33] Maintaining consistency between different types of transactions as much as possible has the advantage of bringing with it a degree of familiarity and might reduce or eliminate any disruptive effect on the law. One the other hand, one objection is that this approach might jar with the specific features of the supply of digital content; for example, remedies such as 'repair' and 'replacement' might sounds somewhat unusual when applied to digital content. So trying to maintain a parallel with familiar rules might seem convenient but does not necessarily provide the best solution.

Secondly, new rules not previously found in respect of goods, but which would provide relevant rules for some of the specific issues created by digital content, could be developed. For example, a new app downloaded onto a smartphone can cause other apps to malfunction and lose data, or the digital

---

[28] The definition of goods in the Consumer Guarantees Act 1993 (New Zealand) includes 'to avoid doubt ... computer software' (s. 2(1)(vi)).
[29] Section 34 CRA.
[30] Section 35 CRA.
[31] Section 36 CRA.
[32] Sections 42–44 CRA.
[33] Sections 42(6)/(7)(a) CRA.

content loaded on a physical device might have defect which causes the device to malfunction and damage the device itself or other goods, or even injure someone. Thus, the CRA includes a specific provision dealing with a remedy for damage caused by digital content (s. 46). This section applies where a trader supplies digital content to a consumer under a contract, and that digital content causes damage to a device or other digital content belonging to the consumer. If the damage is of a kind that would not have occurred had trader exercised reasonable care and skill, then the consumer can require the trader to provide a remedy either by repairing the damage, or to compensate the consumer for the damage with an appropriate payment.[34]

Third, special rules might reflect the fact that there are multiple ways in which a person can access digital content: it could be supplied on a physical medium, downloaded via the Internet, or simply accessed online without being installed on any of the user's devices. Again, the CRA serves as an example: sections 39(3)–(7) concern the situation where a contract to supply digital content provides that once the trader has supplied that content, the consumer is to have access to a 'processing facility', defined as 'a facility by which [the trader] or another trader will receive digital content from the consumer and transmit digital content to the consumer (whether or not other features are to be included under the contract)'. Section 39(5) requires that this facility must be available to the consumer for a reasonable time, or for the time specified in the contract.

Overall, it seems that there are a number of key questions which need to be resolved, and these revolve around two primary issues: first, what expectations or requirements as to the quality and fitness for purpose of digital content should be included in appropriate legal rules, and what would be appropriate remedies where digital content fails to achieve this? Secondly, who should be liable in such a situation, and for what? For example, should liability fall solely on the final supplier of the digital content to the user, or should the creator/owner of the software (assuming they are different from the supplier) be subject to liability, whether on a joint or exclusive basis? It may well be that there cannot be a simple answer to this question, because of the manifold ways in which digital content is supplied and utilised.

6.2. THE CONSUMER NOTION – WHEN DOES A CONSUMER BECOME A TRADER?

A second issue which needs to be considered is whether the established bifurcation between consumers and traders can still be maintained in view of the

---

[34] The EU Commission's proposal for a directive on digital content, discussed in section 7.1 below, contains a provision on this issue in Art. 14 which takes a somewhat different approach to this.

fact that private individuals increasingly become suppliers of goods, services and digital content via various online platforms. Thus, sharing economy platforms such as Airbnb or Uber allow individuals to share their resources with others, whereas other platforms such as Shapeways allow individuals to market their CAD files for 3D-printed goods or to sell homemade items (e.g. Etsy). In some of these cases, the supplier of the goods or services will be acting in a professional capacity and this will be obvious, but in other instance, it will be much less apparent whether the supplier is a business or private individual.

However, this issue is becoming increasingly important, because the recipient of the goods or services will often not know whether the contract is with another private individual, or with a professional seller/supplier. As a result, it will not be clear whether consumer law rules will apply to a particular transaction. Similarly, a private individual who regularly offers goods, services or digital content such as CAD files might not know at which point these activities cross the threshold to being a business supply.

This blurring of the dividing line between when a supply would be done as a business or as a private individual suggests two possible responses: first, it might be necessary to clarify when that diving line is crossed and an individual should be treated as acting in the capacity of a business. In EU law, the familiar definition of a trader, as found for example in Article 2(2) of the Consumer Rights Directive, is ' any natural person or any legal person … who is acting … for *purposes relating to his trade, business, craft or profession*'. However, what are the criteria by which it is determined what constitutes an individual's trade, business, craft or profession, particularly in the case of a 'hobbyist' who has started to offer something occasionally but does so with increasing frequency? It might be necessary to supplement the current definition with an indicative set of factors to be taken into account, such as the regularity with which an individual offers something, the variety of goods or digital content offered, the volume of transactions concluded by an individual, the extent to which that individual profits from the transaction beyond recovering any expenses incurred, and the sophistication of the individual. These are merely suggestions – other factors might also be relevant, and some of those mentioned might not be particularly helpful. For example, in the context of the sharing, or collaborative, economy, the European Commission has suggested that relevant factors might be the frequency with which a service is provided, whether this is done with a profit-seeking motive (rather than cost compensation), and the provider's level of turnover from the same activity.[35]

Secondly, it is also possible to approach this issue from a different angle. As mentioned, this question arises in the context of online platforms, so in the context of a specific category of business model. That being the case, it might be

---

[35] European Commission, *Communication on a European Agenda for the Collaborative Economy*, COM(2016) 356 final, pp. 9–10.

better to consider creating specific rules for this, rather than simply trying to extend the scope of rules designed for a different context. The dominant concern here is whether the recipient of the goods, services or digital content should be entitled to a comparable level of consumer protection as would be the case under a simple contract between a trader and a consumer. If it is assumed that this should be so, then some thought needs to be given to how best to achieve this. It could be done, as suggested above, by facilitating the application of existing consumer law rules through an enhanced definition of key terms. However, it might be better to consider whether some liability should be imposed on the intermediary, i.e. the platform, instead, or whether some form of mandatory insurance to protect the recipient would be a suitable alternative means of ensuring adequate protection for the recipient. The role of platforms and intermediaries raises additional questions which are considered next.

6.3. PLATFORMS AND INTERMEDIARIES

The discussion so far as already noted the emergence of a new type of online intermediary, usually referred to as 'online platforms'. The essential function of such platforms is to act as an intermediary between a supplier of goods, services or digital content and a recipient of the same. As already suggested, such a supplier could be a private individual offering the odd thing on an occasional basis (whether that is a stay in a spare bedroom or a car-share on a specific journey, or handmade craft items) as well as a self-employed trader doing so on a regular basis (e.g. a driver of a private hire vehicle). The purpose of such platforms is to put suppliers and recipients in touch with one another, often for the benefit of the supplier who will gain access to a much wider range of potential customers without having to invest in setting up a personal website. Such platforms may seem to be a modern version of noticeboards in a public library or bulletin boards. However, often these platforms are more than just passive facilitators between the supplier and recipient of goods, services or digital content: some platforms stipulate quite detailed rules which both suppliers and recipients must adhere to. Moreover, many platforms also provide mechanisms for processing orders and payments, taking a small percentage of the price paid as a processing fee. Other platforms may take an even more active role, such as some of the platforms dealing with 3D printing: such a platform may allow a designer to publish their designs (i.e. their CAD files) on the platform and create 'shops', which then allow interested recipients to view the designs and order the physical item. The platform will then process the order, 3D print the item, and despatch the physical item directly to the recipient.

The fact that these platforms are often more than just passive intermediaries raises new legal questions about the extent to which liability should be placed on them, rather than a supplier, when something goes wrong. Depending on the precise contractual arrangements between recipient, supplier and platform, there may already be a clear allocation of liabilities, but this might not always produce the right balance between all the parties concerned, and so it may be necessary to consider whether 'non-passive' intermediaries should be made subject to specific legal duties towards both the supplier and the recipient in respect of the proper performance of transactions concluded via such platforms. Resolving this issue will require a careful analysis of how such platforms operate, what particular problems can be identified, and the extent to which existing laws can be deployed to address these. For example, a platform which provides payment processing facilities will have some liability for incorrect processing of payments, but it will be less clear if the platform can be required to reimburse payments if the supplier of the goods or services fails to perform the contract as agreed. Moreover, a platform which has created a strong brand image associated with the provision of a particular service might appear to a customer as being the actual trader providing that service, which might suggest that rules which either require greater transparency about the role of the platform and/or which impose some liability in respect of the performance of the contract on the platform.

6.4. COMBINATION OF SALE, SERVICE AND DIGITAL CONTENT IN ONE TRANSACTION BUT DIFFERENT PARTIES

The involvement of platforms is just one instance where a transaction seems to involve three parties, with the intermediary no longer merely being a conduit for the dealings between a supplier and a recipient of goods/services/digital content. There are other instances where three-party situations arise: for example, there may be digital content on a physical item, such as software which controls various functions of a car, or an app on a wrist-device which monitors health data of its user. When a person acquires that device, there may be a contract for the supply of the physical item including the digital content, but then a separate contract with the supplier of the digital content for monitoring use of the item and updating the software or for processing the data generated by the content.

The point is that the long-held paradigm transaction of the one-shot trader–consumer contract is on increasingly shaky ground as these multi-party relationships are becoming more and more common. It is a trite observation that contract law is essentially based around a two-party paradigm (i.e. a 'synallagmatic contract'). Thus far, it has not been causing much difficulty to apply this approach to three-party (or even multi-party) situations where each party has a distinct role to play. Thus, an in-store contract to buy a microwave paid for by the buyer using a credit card is a three-party situation (at least), but there are separate – albeit connected – contracts involved. Moreover, these contracts are generally performed relatively swiftly and have a clear end-point (ownership in the goods is transferred, the obligation to pay is satisfied by payment authorised by the credit-card company, and the trader will receive funds from the credit-card company).

So what is different now? The prominence of digital content certainly has the potential to affect the way in which traditional contract rules might be applied: whilst there is no great difference as far as the sale of the physical item is concerned, the contract as it pertains to the digital content is likely to be a long-term contract with continuous performance. Moreover, if there is an element of non-performance in the digital content contract, then this could have an effect on the quality and fitness for purpose of the physical item too, and so questions will arise as to the distribution of liabilities. There is a question as to whether contract law rules designed for the synallagmatic paradigm can cope with these more complex situations. Ultimately, this might well turn out to be the case, and there might be no cause for alarm. However, this is something which will require close analysis. It certainly seems to be the case that, in the context of the Internet of things, the focus of the law on the physical devices will have to give way to recognising that the main functionality of these devices does not rest with the physical item but instead is largely transferred to the digital content and the data that is collected, transferred and processed. In her useful analysis of the legal implications of the Internet of things, Wendehorst notes that:

> 'the emergence of the [Internet of things] is a serious challenge for traditional rules of contract and contractual liability. Some of the problems encountered are merely gradual in nature. There are, however, some more fundamental issues as smart devices and IoT call into question the very notion of sale and ownership ... the future lies in "device as service"...'[36]

Her observations reinforce the point made above that whilst some aspects of the digital revolution can be dealt with by adjustments to the law, others will be much more fundamental and necessitate a conceptual rethink. The paradigms on which much of the law is based may be displaced by the rise of the Internet of things – and the law truly disrupted.

6.5. THE INTERNET OF THINGS AND AUTOMATED CONTRACTING

As noted above, a further significant development brought about by the digital revolution is the 'Internet of things'. This involves various devices all connected to the Internet capable of exchanging data with one another.

---

[36] C. WENDEHORST, 'Consumer Contracts and the Internet of Things' in R. SCHULZE and D. STAUDENMAYER (eds.), *Digital Revolution – Challenges for Contract Law*, Nomos, Baden-Baden 2016.

This can have basic applications such as utilising a smartphone to control the central heating, lighting and digital TV recorder. However, these devices can also collect data about usage and performance, which can be shared with other devices and processed in order to analyse recurring patterns. This could enable such connected devices to adopt a degree of predictive behaviour, for example switching on the heating or lights at a time when the user would usually do so manually.

One possible application for the Internet of things is both domestic and commercial stock-keeping. For example, digital content operating a refrigerator might be able to keep track of the use-by date of food stored in the refrigerator and remind the user to replace out-of-date items. A somewhat fanciful idea is that it might be possible for the refrigerator to place an order for fresh food directly with a grocery delivery service without any intervention by the user. A more realistic and likely application is in a commercial setting: a store could monitor its stock electronically by keeping track of overall stock levels and collect data on the turnover of its stock. Based on this data, orders could be placed automatically with suppliers at times and for quantities as necessary.

There are two broad implications for the development of the law here: first, the possibility to mine data which can then processed both by the devices themselves as well as being shared with others over the Internet raises questions as to how such data can be controlled (more in the next section). There might also be implications if the processing of data produces unexpected or incorrect results.

Secondly, the possibility for devices to place and accept orders without any human intervention creates potential issues for contract law. At a basic level, there will be the question whether such contracts are valid at all. On this point, Article 12 of the UN Convention on the Use of Electronic Communications in International Contracts is interesting: it provides that contracts concluded through the 'interaction of automated message systems' should not be regarded as invalid or unenforceable solely because there was no review or intervention by a natural person in this process. On the one hand, this is a useful enabling provision for automated contracting, but on the other, it might raise the question of what is meant by 'contract' in view of the lack of any true expression of intention by, and agreement between, the parties. Finding legal rules to deal with particular issues might seem easy, but at the same time, this could disrupt our understanding of the very nature of contract law.

Beyond questions of basic validity and enforceability, there are other issues which are likely to require new legal rules. For example, if a device misinterprets the data it has collected and places an incorrect order, who should be liable for the consequences of this: should it be the owner of the device which placed the order, or should such liability potentially be imposed on the supplier of the device, or the supplier of the digital content which processed the data and placed the order? These are questions that will need to be addressed in due course, not least to provide clarity for consumers and traders alike.

6.6. THE USE OF PERSONAL DATA

Finally, one of the most problematic aspects of the digital revolution, and one which where the most disruptive effects might yet be felt, is with regard to personal data. It has already been noted several times that one of the effects of the digital revolution is the possibility to collect, mine and process vast amounts of data about an individual's habits as well as the performance of devices. This data can be of great value to a range of persons: businesses might utilise this data to target advertising and promotional offers with greater focus to attract more business. Also, by receiving data about the performance of individual devices, it becomes possible to get a much better understanding of how devices fare for some time after they have been sold. It might even be possible to detect unusual or unexpected ways in which consumers use such devices.

Inevitably, this vast flow of data raises several legal issues. First, there is the need to ensure that those collecting and processing data ensure that this data is kept securely and only used for authorised purposes. Data protection legislation needs to set clear rules about what is and is not permissible, and these rules need to be enforced rigorously. Secondly, it is also important to recognise the benefits that could arise from data collection and analysis, particularly with regard to usage and performance data. For manufacturers, such data can provide important information about the reliability and durability of their products, as well as about product aspects which could be improved. For this to happen, it must be possible for such data to flow freely without being impeded by jurisdictional limitations. In turn, one might consider whether the right to receive and process such data should also bring with it additional obligations, such as a duty to warn users if common problems or misuses have been identified. Moreover, identification of a particular problem might trigger a duty to modify the product, or, in the case of digital content, a duty to update that content to remove any problems.[37]

A further aspect of data has also started to attract the attention of scholars and policy-makers: an alternative consideration, or 'counter-performance', to money.[38] Instead of paying for digital content such as a mobile phone app, the recipient agrees to make available to the supplier certain personal data. Consumers may not realise that digital content offered for free is, in fact,

---

[37] This is explored in depth by B. WALKER-SMITH, 'Proximity-Driven Liability' (2014) 102 *Georgetown Law Journal* 1777.
[38] C. LANGHANKE and M. SCHMIDT-KESSEL, 'Consumer Data as Consideration' (2015) *Journal of European Consumer and Market Law* 218.

frequently supplied in return for access to some personal data. If it is recognised in law that using data can constitute consideration, or a counter-performance, then such a transaction will also give rise to a contract. There are several issues arising from this. First, personal data is subject to data protection legislation, and it is not usually possible to force someone to permit another person to have access to personal data. This could be a problem where a trader requires personal data to be made available before digital content is supplied because there might be a conflict with data protection rules regarding consent to data processing.

Secondly, assuming a person consents to the provision of personal data to a trader in return for the supply of digital content, under data protection laws this consent can be withdrawn at any time.[39] However, this could create a problem if personal data was provided as consideration for the receipt of digital content, or even goods or services, because there is no clarity as to what the effect of the withdrawal of consent might be on the supply contract. The withdrawal of consent to processing personal data requires the recipient of that data to return it and/or delete the information. As a result, it may seem that the trader no longer has any consideration for the supply of the digital content, so how will this affect the contract itself?

There are a number of possible answers to this. One plausible analysis would be to say that if a contract involves the provision of personal data as consideration, then this could be treated as a licence to use the data in return for the licence to use the digital content, so by withdrawing the consent for processing data and thereby terminating the licence to use the data, the licence to use the digital content is equally terminated immediately. This might be a suitable outcome. An alternative analysis would depend on the extent to which the personal data belonging to one customer has been combined with the personal data of others and processed to produce new data, i.e. where additional value has been derived from an individual's personal data through combining it with other data and analysing this. At this point, even though the personal data supplied by the individual who has now withdrawn his consent is no longer available to the trader for further processing, the analytical findings already obtained presumably are beyond the reach of the individual's consent. In that case, the trader has retained some benefit of value and the individual should be allowed to continue to use the digital content even though consent to data processing has since been withdrawn.

As well as this individual dimension of using data as a form of counter-performance to receive goods, digital content or services, the collection of vast amounts of data are turning data itself into a new subject for business

---

[39] Cf. Arts. 6(1)(a) and 9(2), and Art. 17 of Regulation 2016/679 on the protection of natural persons with regard to the processing of personal data and on the free movement of such data [2016] OJ L119/1 (the new 'Data Protection Regulation').

transaction, and a potentially very valuable one at that.[40] Utilising data in this way brings with it familiar concerns about data security and protection of personal data. However, it also prompts interesting questions about whether data as a commodity, i.e. subject matter of a contract, requires a new legal response.

6.7. OTHER QUESTIONS

The foregoing sections have highlighted what are perhaps the most immediate and significant questions from a legal perspective. There are others, some of which will in due course attract the attention of the courts, regulators or legislatures. For example, Cifrino considers how the ownership of virtual assets within virtual worlds (he focuses on World of Warcraft and Second Life) should be dealt with by law, i.e. whether this should be a matter for property law (which is traditionally concerned with tangible assets or land) or contract law. He concludes that contract law, rather than property law, should resolve matters arising from disputes over virtual assets, with the end-user license agreements (EULAs) applicable to such virtual worlds defining the dispute-resolution process.[41]

The inevitable consequence of a disruptive development is that the full extent of the changes it brings about will not be immediately obvious, but over time clarity on this will emerge. Some implications will become obvious sooner than others, and will attract the attention of legal scholars and policy-makers alike. Those discussed above are already receiving a great deal of attention, with the EU having made the Digital Agenda a priority objective.

7. THE EU AND THE DIGITAL REVOLUTION

This section will consider how the European Union is responding to the challenges of the digital revolution. It is tackling this in stages. A number of initiatives were launched under its 'Digital Single Market' banner,[42] with early proposals for directives in the consumer contract law field (discussed next). At the end of May 2016, the European Commission presented its next wave of proposals and communications. In view of what was said above about the need to plan carefully before any new legal rules are adopted, it is reassuring that the European Commission does not intend to rush towards adopting regulatory

---

[40] A. DE FRANCESCHI and M. LEHMANN, 'Data as Tradeable Commodity and New Measures for their Protection' (2015) 1 *Italian Law Journal* 51.
[41] C.J. CIFRINO, 'Virtual property, virtual rights: why Contract Law, not Property Law, must be the governing paradigm in the law of virtual worlds' (2014) 55 *Boston College Law Review* 235.
[42] The cornerstone document is the European Commission's communication *A Digital Single Market for Europe,* COM(2015) 192 final.

measures for online platforms. In its recent communication on this issue, it states that:

> 'the need to foster the innovation-promoting role of platforms requires that any future regulatory measures proposed at EU level only address clearly identified problems relating to a specific type or activity of online platforms in line with better regulation principles. Such problem-driven approach should begin with an evaluation of whether the existing framework is still appropriate.'[43]

There are two important points to take from this: (i) the starting point will be the *clear* identification of problems, so any action taken will be responsive rather than predictive; and (ii) before new rules are introduced, the application of existing rules to newly identified problems will be considered first. However it is also acknowledged that this may not always be feasible:

> 'The collaborative economy is a good example where rules designed with traditional and often local service provision in mind may impede online platform business models. This issue will be addressed in the forthcoming Commission Communication on the collaborative economy.'[44]

It is reassuring to see that the European Commission has recognised the importance of approaching the challenges created by the digital revolution in this way. This approach is sensible and it will be interesting to see how it is applied in practice. There appears to be a recognition that it might be possible to respond to any identified problems without having to adopt dedicated legislation,[45] perhaps by clarification of how existing rules should be applied in the context of platforms. That said, it does not rule out new regulatory measures, should this prove to be necessary.

7.1. PROPOSALS ON CONTRACT LAW ASPECTS

For present purposes, the most immediately relevant are two proposals put forward by the European Commission in December 2015, one focusing on fully harmonised rules for the online and distance sale of consumer goods,[46]

---

[43]   European Commission, *Communication on Online Platforms and the Digital Single Market – Opportunities and Challenges for Europe*, COM(2016) 288/2, p. 5.

[44]   *Ibid.*

[45]   For a different position, see C. BUSCH, H. SCHULTE-NÖLKE, A. WIEWIÓROWSKA-DOMAGALSKA and F. ZOLL, 'The Rise of the Platform Economy: A New Challenge for EU Consumer Law?' (2016) 5 *Journal of European Consumer and Market Law* 3, calling for work on a 'platform directive' to commence. The Commission's more cautious approach seems preferable at this point.

[46]   Proposal for a Directive of the European Parliament and of the Council on certain aspects concerning contracts for the online and other distance sales of goods, COM(2015) 635 final.

and, perhaps more significantly, the second seeking to introduce new rules for the supply of digital content.[47] Rather than discussing the substance of these proposals here (not least because they are likely to undergo modification during the legislative process),[48] a number of general observations are made instead.

First, these proposals only deal with one of the issues identified in section 6 above: the possible regulation of contracts for the supply of digital content. Indeed, the proposal for the online and distance sales of goods seems to have no real significance as it overlaps with the Consumer Rights Directive (2011/83/EU) and the Consumer Sales Directive (99/44/EC), and it primarily seems to have the more mundane objective of updating the legal rules on what is now a fairly established method of selling goods. Whether this proposal contains any meaningful improvements which would genuinely make the online/distance sale of goods easier throughout the Single Market is for discussion elsewhere, except to note that one cannot help but notice that the overriding concern is to introduce a set of fully harmonised rules rather than necessarily a set of rules clearly targeted at the particular features of such contracts.

Secondly, it is important to appreciate that many of the rules in both proposals have had a lengthy gestation period. They are taken from the proposal for a Common European Sales Law (CESL),[49] which met with a rather lukewarm reception and was eventually withdrawn. The proposal for CESL was the culmination of a decade-long process on developing a European Contract Law,[50] which had resulted in the so-called Draft Common Frame of Reference.[51] So, in a sense, the two proposals put forward in December 2015 are the latest attempt to adopt legislation based on years of work. On the one hand, drawing on previous initiatives is understandable, but on the other, it needs to be questioned to what extent the model rules from the DCFR or its modified provisions in the CESL are actually suitable for dealing with the particular features of contracts for the supply of digital content.[52] Of course,

---

[47] Proposal for a Directive of the European Parliament and of the Council on certain aspects concerning contracts for the supply of digital content, COM(2015) 634 final.
[48] Moreover, several of the other chapters in this collection focus on aspects of these proposals.
[49] COM(2011) 635 final. For scholarly discussions, see e.g. G. DANNEMANN and S. VOGENAUER (eds.), *The Common European Sales Law in Context*, Oxford University Press, Oxford 2013; M. SCHMIDT-KESSEL (ed.), *Ein einheitliches europäisches Kaufrecht?* Sellier, Munich 2012; H. SCHULTE-NÖLKE, F. ZOLL, N. JANSEN and R. SCHULZE (eds.), *Der Entwurf für ein optionales europäisches Kaufrecht*, Sellier, Munich 2012.
[50] Generally, see L. MILLER, *The Emergence of EU Contract Law – Exploring Europeanization*, Oxford University Press, Oxford 2011 or C. TWIGG-FLESNER, *The Europeanisation of Contract Law*, 2nd ed., Routledge, Abingdon 2013.
[51] STUDY GROUP ON A EUROPEAN CIVIL CODE/RESEARCH GROUP ON THE EXISTING EC PRIVATE LAW (ACQUIS GROUP) *Principles, Definitions and Model Rules on European Private Law – Draft Common Frame of Reference*, Oxford University Press, Oxford 2010.
[52] Cf. ARNERSTÅL, *supra* n. 26.

the digital content proposal does contain some modifications. For example, the proposed 'conformity with the contract' requirement in Article 6 prioritises the subjective agreement of the parties in the contract (see also recital 25), and only stipulates a 'fitness for normal use' default criterion if there is no agreement. This raises two issues: first, it invites the question of what the purpose of a 'conformity' requirement in modern consumer legislation should be: is it simply to give effect to the intention of the parties, or is it more ambitious and seeks to set down a clear quality standard which any digital content supplied to consumers must meet? Bearing in mind the objective of creating a digital single market throughout the EU, it seems that a clear objective quality standard would be better suited for this. Secondly, and following on from the previous point, the use of the 'conformity with the contract' notion and the priority given to the agreement of the parties itself is open to challenge for the simple reason that it does not match the realities of acquiring digital content: surely in most cases, the process of acquiring digital content is fully automated, with no negotiation between the parties and therefore no agreement reflected in the contract, and little, if any, human intervention on the trader's side?[53] So a subjective conformity requirement seems not only a backwards step but one which is simply unsuitable for the nature of the transaction. A novel solution reflecting the specific features of digital content and the supply of such content is needed.

7.2. OTHER PROPOSALS

In addition to the proposals mentioned in the previous section, there are further proposals which relate to the digital single market and have at least some impact on contract law. For example, the proposed 'Geo-Blocking' Regulation[54] is intended to prevent discrimination against customers by traders on the grounds of the customer's residence when accessing websites or ordering goods or services. This is essentially a specific application of the EU's non-discrimination principle which is already a limitation to the scope of the general principle of freedom of contract. A further proposal deals with cross-border parcel delivery services,[55] but this only strengthens the oversight of the relevant regulators and requires universal service providers to notify their tariffs, which are then

---

[53] Cf. J. SMITS, *The New EU Proposal for Harmonised Rules for the Online Sales of Tangible Goods: Conformity, Lack of Conformity and Remedies*, Study for the JURI Committee of the European Parliament (February 2016), p. 9.
[54] Proposal for a Regulation on addressing geo-blocking and other forms of discrimination based on customers' nationality, place of residence or place of establishment within the internal market, COM(2016) 289 final.
[55] Proposal for a Regulation on Cross-border parcel delivery services, COM(2016) 285 final.

assessed for their affordability, as well as dealing with cross-border access. It does not provide any specific contract law rules.[56]

7.3. WHAT NEXT?

Whatever might be said about the detail of the proposal on digital content, the fact that a proposal on contract for the supply of digital content has been made has an obvious signalling function: the disruptive effect of the digital revolution has reached the law of contract and the time to consider the implications flowing from this has come. Designing appropriate rules for the supply of digital content will only be the start. There are many issues which will require further consideration (several of which are already on the EU's to-do list). A lingering issue will undoubtedly be the role of online platforms and whether new rules will eventually be considered for these. Currently, there is no specific EU law in place, nor envisaged, which focuses specifically on these platforms. Existing rules in Articles 12–14 of the E-Commerce Directive (2000/31/EC) deal with the liability of intermediaries in the context of e-commerce (which includes such platforms), but that liability is very limited.[57] As noted at the start of this section, the European Commission intends to proceed in a problem-focused manner and investigate first whether existing legal rules can address newly identified problems with platforms before developing new rules. Whilst this seems to entail maintaining the current position regarding the liability of platforms under the E-Commerce Directive, there will be targeted intervention dealing with content harmful to minors and hate-speech as well as the allocation of revenues from the distribution of copyright-protected content.[58] The main focus for the EU with regard to online platforms is to create a regulatory environment that facilitates innovation, and as such, top-down regulation is not on the cards immediately. With existing rules such as the Unfair Commercial Practices Directive (2005/29/EU) and the Unfair Contract Terms Directive (93/13/EEC) already creating a strong consumer protection framework, ensuring compliance and full enforcement of existing rules seems an appropriate response at this time. Indeed, it its Communication on the collaborative economy,[59] the European Commission sets out its thinking about collaborative economy platforms. To the extent that such platforms are only providing an 'information society service'

---

[56] There is also a proposal for an updated version of the consumer protection enforcement regulation: Proposal for a Regulation on cooperation between national authorities responsible for the enforcement of consumer protection laws, COM(2016) 283 final.

[57] Cf. Case C-324/09, *L'Oréal SA and Others v eBay International AG and Others* [2011] ECR I-6011.

[58] European Commission, *Communication on Online Platforms and the Digital Single Market – Opportunities and Challenges for Europe*, COM(2016) 288/2, p. 9.

[59] European Commission, *Communication on a European Agenda for the Collaborative Economy*, COM(2016) 356 final.

within the meaning of the E-Commerce Directive, their liability will continue to be limited, but if a platform is also treated as providing the underlying service, additional obligations such as licensing requirements will apply. The Commission acknowledges that whether a particular platform should be treated as providing the underlying service requires a case-by-case analysis, with relevant factors including the extent to which the platform sets the final price and the key contract terms for the contractual relationship between recipient and provider, and whether the platform owns the key assets used in providing the service.[60] The Communication also considers questions relevant to establishing the employment status of a collaborative economy service provider and taxation issues.[61] The key point to note about this Communication is that it adopts a cautious approach, primarily by seeking to clarify the main issues currently affecting the sharing economy and – at least for the time being – by setting out how existing legal rules could be applied to these particular issues.

There is also the greater prevalence of connected devices in the Internet of things and the wealth of data created and exchanged as a result, as well as the legal implications of fully automated contracting. Similarly, the way data is being used both as a form of 'counter-performance' and as a commodity in its own right raises new issues. For all of these areas, and others which might also be targets for EU intervention, the key message this chapter has sought to convey is that care needs to be taken not to be overly hasty and that any concerns about filling gaps in the law quickly should not drive the EU's legislature to adopt rules which might be just 'good enough' – this is not an area of law where rules for rules' sake are needed. Rather, the issues and legal implications of these developments and their disruptive effects for the law need to be identified clearly, and possible ways of responding to these developed and scrutinised before further proposals are made.

8. A CONCLUDING THOUGHT

There can be no doubt that the digital revolution has been a disruptive technology for business, but the disruptive effect for law, especially contract law, is not necessarily as extensive as the flurry of activity at the EU level might suggest – although the careful approach set out in the Commission's communication on online platforms discussed above seems to be the right one at this point in time.

Generally speaking, it is essential to consider the implications for the law of the disruptive effects of the digital revolution carefully and to identify where

---

[60] Ibid., pp. 6–8.
[61] It was noted earlier (see section 6.2) how the Commission thinks the existing definition of 'trader' could be applied in determining when a private individual crosses the threshold of that definition.

there is a true disruptive effect on the law which will necessitate the development of new legal rules. As well as finding solutions for such specific issues, the wider implications for the affected areas of law as a whole need to be considered before new rules are adopted. Disruptive technology poses challenges for the law and can result in disrupted law, and the task for legal scholars and policy-makers is to consider appropriate solutions. One of the factors affecting the ability of a business to adapt to disruptive technology identified by Christensen in the conclusions to his book is that the capabilities of an organisation are often specialised and context-specific. He writes:

> 'All of these capabilities – of organizations and individuals – are defined and refined by the types of problems tackled in the past, the nature of which has also been shaped by the characteristics of the value networks in which the organizations and individuals have historically competed. Very often, the new markets enabled by disruptive technologies require very different capabilities…'[62]

A similar observation could be made about the law: its rules and principles were defined and refined by problems which have arisen in the past, and this has shaped the value system on which legal rules are based. However, law which has been disrupted requires new rules and principles to deal with the novel issues which have been created by the effects of disruptive technology. The challenge for legal scholars, policy-makers and legislators alike is to identify what these new rules, principles and concepts need to be, and rising to this challenge will require new capabilities to be developed.

---

[62] CHRISTENSEN, *supra* n. 2, p. 227.

PART II

DATA AS A TRADEABLE COMMODITY AND THE NEW INSTRUMENTS FOR THEIR PROTECTION

DATA AS A TRADEABLE COMMODITY

Herbert ZECH*

1. Data as the Object of a Contract 53
   1.1. Defining Data 53
        1.1.1. Data as Machine-Readable Encoded Information 53
        1.1.2. Semantic, Syntactic and Structural Definition 53
        1.1.3. Definition of 'Digital Content' under EU Law 55
   1.2. Data as Something Useful 56
        1.2.1. Possessing Data: Access 56
        1.2.2. Using Data 57
        1.2.3. Destroying Data: Integrity 57
   1.3. Data Markets: Value Chains in Big Data Scenarios 57
        1.3.1. Production, Collection and Analysis of Data 57
        1.3.2. Companies and Consumers as Parts of the Value Chain ... 58
   1.4. Trading Data: Factual and Legal Positions 59
        1.4.1. Having Data: Access, Sole Access (Secret Data) –
               Factual Position 59
        1.4.2. Owning Data: Exclusive Right – Legal Position
               (as Property Assignable) 59
        1.4.3. Non-Tradeable Positions 60
   1.5. The Role of Contract Law, IP Law and Data Protection 60
        1.5.1. Contract Law: Terms and Conditions, Fairness,
               Binding *Inter Partes* 60
        1.5.2. IP Law: Legal Exclusivity, Transferable/Assignable,
               Erga Omnes (Rights to Use Data) 61
        1.5.3. Data Protection: Non-Assignable, *Erga Omnes* 61
2. Existing Exclusive Rights for Different Kinds of Data 62
   2.1. Secret Data: Factual Exclusivity and Trade Secret Protection 62
        2.1.1. Data as Trade Secrets 62
        2.1.2. Partial Property-like Protection 63

---

* This article draws partly on previous publications: 'Information as Property' (2015) 6(3) *Journal of Intellectual Property, Information Technology and Electronic Commerce Law* 192, and 'A legal framework for a data economy in the European Digital Single Market: rights to use data' (2016) *Journal of Intellectual Property Law & Practice*, forthcoming.

		2.1.3. Difficult Determination of the Protected Person 64
		2.1.4. Complex Contracting 64
	2.2. Technical Protection Measures: Factual Exclusivity and Legal
		Protection 65
		2.2.1. Factual Exclusivity without Secrecy 65
		2.2.2. Legal Protection. 65
		2.2.3. Contracts Concerning Protected Data. 65
	2.3. Personal Data: Data Protection 66
		2.3.1. No Comprehensive Allocation of Data to the Person
			Concerned 67
		2.3.2. Property-like Allocation of Personal Data? 67
		2.3.3. Is Data Protection as a Property Justified? 69
	2.4. Databases: Right for the Maker of a Database 69
		2.4.1. Collections of Data as Databases 70
			2.4.1.1. Collection 71
			2.4.1.2. Of Independent Works, Data or Other Materials . 71
			2.4.1.3. Systematic or Methodical Arrangement 72
		2.4.2. Protection Requirement of Substantial Investment. 72
		2.4.3. No Exclusive Right in Individual Data 73
3. Introduction of a Data Producer Right? 74
	3.1. Features of a Data Producer Right. 74
		3.1.1. Subject Matter 74
		3.1.2. Protection Requirements 74
		3.1.3. First Ownership (Original Rights Holder) 75
		3.1.4. Scope of Protection, Exceptions and Limitations 75
		3.1.5. Transferability 76
		3.1.6. Interaction with Existing Rights. 76
	3.2. Possible Justification: Creation of a Market for Data (Fairness
		Aspects) 76
		3.2.1. Incentives for Generating and Disclosing Data 77
		3.2.2. Creating a Market for Raw Data in Big Data Scenarios 77
		3.2.3. Clear Allocation of First Ownership 78
	3.3. Concerns: Free Flow of Data. 78
4. Conclusion. 79

Transactions concerning data challenge both contract law and property law. The Digital Single Market Strategy[1] addresses data protection and copyright but not contract law. The proposed Directive concerning contracts for the supply of

---

[1] Communication from the Commission to the European Parliament, the Council, the European Economic and Social Committee and the Committee of the Regions, A Digital Single Market Strategy for Europe, COM(2015) 192 final.

digital content[2] fills this gap but only applies to business to consumer contracts and is silent about the qualification of data contracts. This article seeks to clarify the factual and legal situation underlying contracts concerning data. First, the concept of data as a commodity and the interplay between contracts, factual exclusivity and legal exclusivity is discussed. Second, the legal framework for contracts concerning different kinds of data is shown. Data as a commodity and their use may be defined mainly by contractual provisions. This is the case when contracts concern secret data. In business-to-business relations this solution is identical to contracts concerning trade secrets. Similarly, data may be traded using technical protection measures leading to a factual exclusivity of their use. Data protection, originally being an aspect of personality protection, may be used as an object of legal transactions too - but only to a certain extent. Within existing intellectual property rights, the right for the maker of a database is licensable and transferrable but does not protect the single datum. This leads to the question whether a new kind of data property right should be created. Therefore, third, the possible outline and arguments in favour of and against the introduction of a data producer right are presented.

1. DATA AS THE OBJECT OF A CONTRACT

When data are discussed as a legal object, a clear definition is not always used. The term digital content as defined by European law may be used as an alternative. As a prerequisite for the analysis of contracts and property rights related to data, the terminology has to be clarified.

1.1. DEFINING DATA

1.1.1. *Data as Machine-Readable Encoded Information*

In its simplest meaning, the term data can be defined as machine-readable encoded information.[3] For contracts and property rights, however, data have to be defined as a legal object or as an economic good.

1.1.2. *Semantic, Syntactic and Structural Definition*

Semiotics demonstrates the exceptional importance of signs representing information.[4] The semiotic distinction between the semantic level of

---

[2] Proposal for a Directive of the European Parliament and of the Council on certain aspects concerning contracts for the supply of digital content, COM(2015) 634 final.
[3] H. ZECH, *Information als Schutzgegenstand*, Mohr Siebeck, Tübingen 2012, p. 32.
[4] U. ECO, *A Theory of Semiotics*, Indiana University Press, Bloomington 1978.

information (meaning), the syntactic level of information (signs and their relation with each other) and the communication channel (on the physical level) leads to the distinction between the content layer, code layer and physical layer. When talking about information transfer in modern information technology the distinction between these three layers as proposed by Benkler and Lessig[5] plays a crucial role. In the discussion about big data therefore it is very important to distinguish between 'raw' data and actual knowledge.[6]

The distinction between the content layer, code layer and physical layer provides a powerful tool for defining information that can be treated as an object: it reveals that information can be defined on the semantic level (information with a certain meaning), on the syntactic level (information represented by a certain amount of signs), or even by its physical carrier (information contained in a certain physical carrier or in a wider sense information represented by the structure of a physical object). Each of the three types of information can be found in everyday life – when we talk about the news, a story or the 'content' of a book we refer to the semantic level. Handling a text or a file refers to the syntactic level. Finally, dealing with a CD, a printed book, etc. refers to the structural level. The three levels are of course connected as meaning can be contained within a text and a text can be printed. Thus, the physical layer carries the syntactic layer and the syntactic layer the semantic layer. Nevertheless, from an economic and from a legal perspective, each layer represents independent possibilities to define a certain amount of information.

Accordingly, data can be defined on each of the three layers. Trading data can mean either trading data with a certain meaning or content, or trading data as an amount of signs (especially digitally recorded like a file or a data stream), or – as was usual before the development of powerful network technology – trading physical carriers containing data like CDs or memory sticks. A distinction on the level of meaning (semantic information) is made in the case of personal data defined as information relating to a person (Article 4(1) General Data Protection Regulation).[7] Likewise know-how, when differentiated by its meaning, is deemed to be semantic information. However, the term data can also be defined on the level of signs (syntactic information), regardless of its meaning. As a legal object, a sequence of 'ones and zeroes' could be protected or traded, either as a file or as a data stream.

---

[5] Y. BENKLER, 'From Consumers to Users: Shifting the Deeper Structures of Regulation Toward Sustainable Commons and User Access' (2000) 52 *Federal Communications Law Journal* 561, 562; L. LESSIG, *The Future of Ideas, The Fate of the Commons in a Connected World*, Random House, New York 2002, p. 23.

[6] N. SILVER, *The Signal and the Noise*, Penguin, New York 2012, p. 13.

[7] Regulation (EU) 2016/679 of the European Parliament and of the Council on the Protection of individuals with regard to the processing of personal data and on free movement of such data, and repealing Directive 95/46/EC (General Data Protection Regulation) [2016] OJ L119/1.

1.1.3. Definition of 'Digital Content' under EU Law

The term 'digital content' is used by European sales contract law. Article 2(10) Directive 2011/83/EU defines digital content as 'data which are produced and supplied in digital form'. The proposed Directive on certain aspects concerning contracts for the supply of digital content[8] contains a definition in Article 2(1):

> '"digital content" means
> (a) data which is produced and supplied in digital form, for example video, audio, applications, digital games and any other software,
> (b) a service allowing the creation, processing or storage of data in digital form, where such data is provided by the consumer, and
> (c) a service allowing sharing of and any other interaction with data in digital form provided by other users of the service ...;'

It seems that the term data is used on the syntactical level ('digital form'). In contrast, the Directive also uses the term 'personal data or any other data' (Art. 3(1), stating that the Directive also – and only – applies if such data are actively provided in exchange for the supply of digital content). The term 'personal data or any other data' is used distinctly from 'data which is produced and supplied in digital form'. At least 'personal data' which refers to data protection law has to be understood in the conventional semantic sense.

The Directive also addresses the problem of differentiating data and their physical carrier. Article 3(3) extends the Directive's applicability to 'to any durable medium incorporating digital content where the durable medium has been used exclusively as carrier of digital content'. This encompasses data supplied on physical carriers but not physical carriers that are not used exclusively for the purpose of supplying data.

There is also a differentiation between digital content and services: according to Article 3(5)(a) of the proposal the Directive shall not apply to contracts regarding services performed with a predominant element of human intervention by the supplier where the digital format is used mainly as a carrier. However, contrary to this distinction, storage services (like cloud services) are treated as digital content according to Article 2(1)(a).

With the CJEU's *UsedSoft* decision[9] (regarding software resales) it became clear that digital goods should be treated as a third category besides (corporeal) goods and services. For IP lawyers who are familiar with licensing contracts

---

[8] Proposal for a Directive on certain aspects concerning contracts for the supply of digital content, COM(2015) 634 final.
[9] Case C-128/11, *UsedSoft v. Oracle* ECLI:EU:C:2012:234; Cf. H. ZECH, 'Vom Buch zur Cloud – Die Verkehrsfähigkeit digitaler Güter' (2013) 5 *ZGE/IPJ (Zeitschrift für Geistiges Eigentum/ Intellectual Property Journal)* 368.

this is not a new insight. The concept of 'digital content' may help clarifying the function of such contracts.

1.2. DATA AS SOMETHING USEFUL

If data are traded as such, they are treated as goods (and not as data-related services). The economic concept of a commodity or good is based on the use aspect. Goods are objects defined by their use. From a legal perspective, prior to analysing the contracts, it has to be asked whether certain goods are subject to property rights. This is the role of legal protection of data.[10] Under Italian law, according to Article 810 *Codice civile* ('*Sono beni le cose che possono formare oggetto di diritti*'), objects of rights (absolute or relative) can be defined as goods.[11] Similarly, in Germany the objects of rights are often identified with goods.

The economic understanding of goods rather looks at the related uses. A good is the sum of its possible uses. This leads to the concept of property rights as a bundle of rights, i.e. a bundle of possible uses assigned to the rights holder. Building on the standard categories of property rights – use (*usus*), enjoying the benefits of the use (*usus fructus*), changing form and substance (*abusus*) and transfer of the property – three basic categories of rights to information can be distinguished: possessing information, using information and destroying information.[12] These categories can be directly translated into categories of data property.

1.2.1. Possessing Data: Access

The first category of information-related activity that can be exclusively attributed to a rights holder is information access. It equals the category of possession in tangible property.[13] Possessing an object enables the owner to perform any kind of activity related to this object, especially to use it. Unlike processing a corporeal object, having access to information is both non-rival and non-exclusive. Therefore, property rights (as well as contracts) regarding access to information should be constructed differently.

---

[10] A. DE FRANCESCHI and M. LEHMANN, 'Data as tradeable commodity and new measures for their protection' (2015) *ItaLJ* (*The Italian Law Journal*) 51, 54.
[11] *Ibid.*, p. 55.
[12] The transfer of a property right is not regarded as a specific category of property like possessing, using or destroying. It belongs to a different level since it is not part of the activities exclusively assigned to the right owner but rather captures the question of the assignability of such a right on a higher level (or meta-level).
[13] Cf. J. RIFKIN, *The Age of Access: The New Culture of Hypercapitalism, Where All of Life Is a Paid-for Experience*, Ken Tarcher/Putnam, New York 2000.

1.2.2. *Using Data*

The second category is information use. Although access to information is a necessary requirement for using information, the two aspects can be attributed differently. An example of this would be the difference between patents and copyright: whereas patents limit the use of information without limiting access (and on the contrary aim at distributing technical information among the public), copyright limits the information by limiting access (namely prohibiting the copying and distributing of copyrighted works). One of the key aspects of data use in big data scenarios is analysing raw data in order to extract new insights.

1.2.3. *Destroying Data: Integrity*

The third category is the destruction of information. This can be achieved by altering syntactic information on the code level or by falsifying semantic information. Moreover, syntactical information can be destroyed completely by deleting it, that is by destroying every existing carrier (structural information) containing the specific syntactic information. Knowledge, that is semantic information in the human mind, cannot be destroyed – or at least it cannot be destroyed without violating the integrity of the persons who have access to it.

1.3. DATA MARKETS: VALUE CHAINS IN BIG DATA SCENARIOS

1.3.1. *Production, Collection and Analysis of Data*

The development of information technology has greatly lowered the cost, size and weight of sensors, memory elements, networks, computers and control elements. Therefore data can be, so to speak, incidentally stored, transmitted and analysed with prospect of profit. In trying to describe the impact of these technologies, three main aspects can be distinguished. Firstly, conventional physical goods start generating, storing and transmitting data. For example, simple cars become 'driving computers'. This situation is often characterised by the addition of the term 'smart' and concerns a broad range of commercial objects and everyday items, such as vehicles, manufacturing machinery, agricultural machinery, mobile phones, home furnishings or clothing (wearables). Secondly, conventional physical goods are increasingly replaced by data (digital content rather than traditional media), for example in the form of e-books, e-papers or streaming services, this especially concerns copyright (and the first pillar of the Digital Single Market Strategy (DSMS)). Thirdly, data is also traded as novel goods of its own kind. Data economy not only means data-driven or data-controlled economy (the control of economic processes through data has existed for a long time and by networking has reached a new quality), but also an economy with data as goods.

Not only are processed data or information traded as economic goods, but also so-called 'raw' or 'machine data'.[14] They are automatically recorded and by analysing large amounts of such data new insights can be gained. This represents the core of so-called Big Data applications. Raw data become a 'commodity'[15] which is traded in a data economy; thus the question of transferable rights of use becomes important. The value chain of these applications can be divided into the production of data (by operation of sensors, either deliberately or incidentally), the collection of data, the analysis by statistical evaluation and ultimately, and – which can only be the last step – innovations based on the resulting insights.[16]

Rights of use or exclusive rights intervene on different levels according to their conditions and their effects. Classic IP (especially patent law, but also copyright) acts only at the level of the innovation process (more precisely at the stage of invention, not at the stage of the upstream idea or the downstream product).[17] The core question in the debate on rights to use data is therefore whether exclusive legal protection should intervene at the stage of data production prior to any innovation.

1.3.2. Companies and Consumers as Parts of the Value Chain

There are two potential groups of actors in a data economy, companies and consumers. Although the state plays an important role, especially in data protection law, its role shall not be further addressed, since special public law provisions exist with the rules on freedom of information (access to information) and further processing of information.[18] The relationship of these rules to potential private sector rights to use data may, however, pose special problems.

Both companies and consumers can be the concerned parties, i.e. the party to which the data relates. If the party is an individual, they are the 'data subject'

---

[14] Term based on MAXIMILIAN BECKER, personal communication.
[15] V. MAYER-SCHÖNBERGER and K. CUKIER, *Big Data – A Revolution That Will Transform How We Live, Work and Think*, John Murray Publisher, London 2013, p. 5.
[16] Indirectly derived innovations can be inventions (technical innovations), copyright protected works (cf. A. WIEBE, 'Der Schutz von Datenbanken – ungeliebtes Stiefkind des Immaterialgüterrechts' (2014) *CR (COMPUTER UND RECHT)* 1, 9) or other innovations. The term innovation is used in the broader sense comprising the whole process leading from ideas to inventions to new products/innovations in the narrower sense.
[17] H. ZECH, 'Life Sciences and Intellectual Property: Technology Law Put to the Test' (2015) 7 *ZGE/IPJ* 1, 3.
[18] See A. WIEBE and E. AHNEFELD, 'Zugang zu und Verwertung von Informationen der öffentlichen Hand' (2015) *CR* 127, and ID. (2015) *CR* 199.

within the meaning of Article 4(1) General Data Protection Regulation.[19] In this case the data is 'personal data'.

Companies and consumers can also act as producers of data. With the proliferation of complex devices, consumers become increasingly important as data producers, whether as operators of 'smart cars'[20] or carriers of 'wearables'. Within the so-called Industrie 4.0 (fourth industrial revolution, internet of things applied to manufacturing) companies become data producers, even if their corporate purpose is aimed at completely different business.

An analysis of the data is regularly done only on the business level, regardless of the source from which the data originates. What seems to be relevant to the intermediary data trade, besides the question of whether or not the data are personal data (and thus data protection law is applicable), is the distinction between consumer-generated and company-generated data. Whether legal consequences for rights to use data should differ depending on whether the generator is a company or a consumer shall be discussed later.

1.4. TRADING DATA: FACTUAL AND LEGAL POSITIONS

1.4.1. Having Data: Access, Sole Access (Secret Data) – Factual Position

Contracts about information often concern only the factual position.[21] Since access is the fundamental position regarding data (instead of possession regarding corporeal things), contracts may be directed to conferring access to certain data. Having been granted access, the other party is able to use the data in further ways like copying, distributing or analysing the data. A special case is that of sole access, which can be conferred as long as the data are kept secret.

1.4.2. Owning Data: Exclusive Right – Legal Position (as Property Assignable)

Data may or may not be allocated by property rights. Since information is a much less clearly defined object than corporeal objects, property rights in information have to be carefully constructed as a bundle of rights. In addition, it should be considered that informational goods – at least semantic and syntactic information – are public goods in the sense that their use is non-exclusive and non-rival. Moreover, information as such is not depreciable, which is especially important for the justification of property rights to information.

---

19   Regulation (EU) 2016/679 of the European Parliament and of the Council on the protection of individuals with regard to the processing of personal data and on the free movement of such data, and repealing Directive 95/46/EC (General Data Protection Regulation), [2016] OJ L119/1.
20   About rights to use data in 'smart cars' see G. HORNUNG and T. GOEBLE, '"Data Ownership" im vernetzten Automobil' (2015) *CR* 265.
21   DE FRANCESCHI and LEHMANN, *supra* n. 10, p. 52 et seq.

1.4.3. Non-Tradeable Positions

The law may hinder interested parties in using data without allocating full property rights. Data protection is the most important example: as a personality right, the function of data protection is not to allocate value. Most importantly, unlike property rights, data protection in general is non-assignable. However, contracts regarding protected data are possible[22] and the question of whether data protection should be regarded as a 'quasi IP' is hotly debated (see section 2.3).

1.5. THE ROLE OF CONTRACT LAW, IP LAW AND DATA PROTECTION

Contracts concerning data may either solely rely on contractual obligations or start with an exclusive right belonging to one party. If the party trading his or her data relies on contractual obligations alone, the resulting legal certainty is limited. Contracts are binding only for the contracting parties. Secret data being disclosed to the other party may be protected as trade secrets but the obligation to keep them secret is only binding for the other party. Should the data be disclosed to the public, third parties cannot be prevented from using the data.

Therefore, factual exclusivity – that is secrecy – is difficult to trade. Although know-how contracts do exist, they require careful drafting and always remain a source of uncertainty. Moreover, for consumers the amount of legal expertise to enter into such contracts is not easily obtainable. Technical protection measures allow a certain degree of factual exclusivity without secrecy. However, again for consumers the use of technical protection measures seems impractical. It is no coincidence that the industry currently sees no great need for the introduction of new exclusive rights in data. From a consumers' perspective – meaning consumers who wish to trade 'their' data – such a right would be more desirable.

Exclusive rights may significantly lower transaction costs – one of the main reasons for the existence of IP rights. Starting from such rights the supplying party may either transfer the right entirely or grant derivative user rights (a solution popular in German copyright law) or simply allow the use of the data despite existing exclusive rights – i.e. grant a licence.

1.5.1. Contract Law: Terms and Conditions, Fairness, Binding Inter Partes

Contracts may stand on their own or rely on existing property rights. Contractual obligations are only binding *inter partes*. Therefore, data have to be kept secret in order to ensure exclusive use by the contracting parties. This may only be achieved by technical protection measures or property rights.

---

[22] Cf. C. LANGHANKE and M. SCHMIDT-KESSEL, 'Consumer Data as Consideration' (2015) *Journal of European Consumer and Market Law* 221.

Contract law also serves an important role in guaranteeing transactional fairness. By providing and ensuring standardised obligations, the law significantly lowers transaction costs.[23] So far there are no standard data contracts. The proposed Directive concerning contracts for the supply of digital content fills this gap, but only applies to business-to-consumer contracts and is silent about the qualification of data contracts.

1.5.2. IP Law: Legal Exclusivity, Transferable/Assignable, Erga Omnes (Rights to Use Data)

Rights to use data or property-like rights concerning data ('ownership') are to be understood as the allocation of data by means or at least along the lines of exclusive rights. Ownership rights as comprehensive rights to tangible property and 'classic' intellectual property rights such as patent law and copyright law serve as exemplary models. They are supplemented by less clearly contoured neighbouring rights, as well as unfair competition law.

In addition to a clear definition of the subject matter, an important criterion concerning the question of whether a property-like allocation of legal powers over an object or good exists is the allocation of economic value. As a consequence, for example, interferences with legal claims result not only in claims for damages, but also give rise to disgorgement of profits based on unjust enrichment and assumed agency and the transferability of rights (which is not necessary for the allocation of economic value – licensability suffices). There is a hierarchical relationship between transferability and allocation of economic value, since the transferability of rights requires the allocation of economic value; however, the allocation of economic value does not presuppose transferability. Rather, it is sufficient for the right to confer the commercial use in other ways than by transferring them completely. At the very least it is required that the subject matter is *de facto* transferable, so that the assigned powers may actually be exercised by a person other than the legal entity, which for example is also the case with certain personality aspects like one's own image (defined semantically).

1.5.3. Data Protection: Non-Assignable, Erga Omnes

The preconditions for regarding personal rights as exclusive rights are highly disputed. The debate mainly concerns the role of data protection and will be discussed in detail in section 2.3.

---

[23] Cf. J. DREXL, 'Zwingendes Recht als Strukturprinzip des Europäischen Verbrauchervertragsrechts' in M. COESTER et al. (eds.), *Privatrecht in Europa*, Festschrift für Sonnenberger, C.H. Beck, Munich 2004, p. 771, 788; M.-O. MACKENRODT, *Technologie statt Vertrag*, Mohr Siebeck, Tübingen 2015, p. 187.

2. EXISTING EXCLUSIVE RIGHTS FOR DIFFERENT KINDS OF DATA

As potential existing exclusive rights regarding data shall be examined trade secret protection, the legal protection of technical protection measures, data protection and the *sui generis* right for the maker of a database.

2.1. SECRET DATA: FACTUAL EXCLUSIVITY AND TRADE SECRET PROTECTION

With the proposal of a Directive on the protection of undisclosed know-how and business information (trade secrets)[24] the protection of trade secrets is the subject of European harmonisation efforts. Although data fall under the definition of trade secrets, the protection does not lead to real exclusivity right regarding data. With regard in particular to Big Data scenarios, this leads to problems.

2.1.1. Data as Trade Secrets

The main protection requirement is that the data constitute a trade secret. This can also apply to automatically collected data, provided that they are related to business matters (for which the storage as part of a business operation is sufficient), that they are non-obvious and that an expressed or at least recognisable intent to maintain secrecy and an economic interest in the confidentiality on the part of the business holder exists. All these conditions can also be met if the manufacturer of a complex machine keeps the data collected by the machine secret from its clients.

With the definition of trade secrets as 'information',[25] there is a demarcation at the semantic level (as in data protection). However, trade secrets are not necessarily information about the protected business (business-related information), but rather can be any kind of information. The only precondition is that the company has the information legitimately under its control ('lawfully within their control', Article 39(2) TRIPS; cf. Article 2(1)(c) of the draft Directive).[26] Articles 3(2)(a) and 11(2)(e) of the draft Directive (and recital 6)

---

[24] Proposal for a Directive of the European Parliament and of the Council on the protection of undisclosed know-how and business information (trade secrets) against their unlawful acquisition, use and disclosure of, COM(2013) 813 final.
[25] Article 2(1) draft Directive. Cf. Article 39(1) TRIPS: 'information'.
[26] In the draft Directive it reads 'trade secret holder', cf. Articles 3, 9, 11 and 13. According to Article 2(1)(c) it is part of the definition of trade secret that the protected information is 'subject to reasonable steps under the circumstances, by the person lawfully in control of the information, to keep it secret'. Article 10(1) requires a legitimate holder as well. The 'legitimate holder' is also mentioned in recitals 3, 5, 6, 13, 14, 15 and 18.

explicitly mention 'electronic files containing or implementing the trade secret' (syntactic information).

According to Article 2(1)(b) of the draft Directive a trade secret must have a commercial value because of its being secret.[27] According to recital 8 the definition should not comprise any 'trivial information'. In the case of a single measurement datum it could be argued that this does not yet have any commercial value and is in any case trivial. This shows that recital 8 is based on false premises: with Big Data trivial information can have economic value when putting together just enough trivial information and analysing it. The existence of a market for such data is likely to disprove the worthlessness of the data. Even raw data have a value which is just very low under certain circumstances.[28] Since the value does not have to exceed any minimum quantitative threshold, the requirement of commercial value should not be a problem. This, of course, also applies to the requirement of commercial interest under German law, for example to manufacturers of complex machines.

2.1.2. Partial Property-like Protection

Although the protection of corporate secrets is considered to be a kind of intellectual property[29] (and with Article 39 falls under the scope of TRIPS), it grants no genuine exclusive right. In particular, it depends on the factual existence of the secret and thus rather resembles the protection of possession.[30] In addition, the information is not protected against all use, but only against certain types of attack on the secret.

In Big Data scenarios the analysis of data by third parties would necessarily imply an infringing disclosure, provided that the third party has no authorised access, for example through the acquisition of data on the basis of a know-how contract. Thus the analysis as a use of secrets that have been obtained by violating secrecy or by commercial espionage would be included in the protection of corporate secrets. It is appropriate that a repeated production of the same data (e.g. by conducting the same measurements once more) remains admissible because the protection does not transfer exclusive powers to the information (incidentally, there are also exclusive rights allowing the independent re-creation

---

27 Cf. GRUR statement of 19.03.2014, 4 et seq.
28 K. NEUMANN, 'Es gibt kein belangloses Datum mehr!' ('There is no trivial datum anymore!') (2011) *DANA (Datenschutz Nachrichten)* 44.
29 See also A. OHLY, 'Der Geheimnisschutz im deutschen Recht: heutiger Stand und Perspektiven' (2014) *GRUR (Gewerblicher Rechtsschutz und Urheberrecht)* 1, 3 et seq.
30 M.H. DORNER, *Know-how-Schutz im Umbruch*, Carl Heymanns, Cologne 2013, p. 111; ID., 'Big Data und "Dateneigentum", Grundfragen des modernen Daten- und Informationshandels' (2014) *CR* 617, 619; OHLY, *supra* n. 29, p. 8.

of an object of protection, cf. Article 19(2) Community Design Regulation[31] for the unregistered Community design).

The legal position of the protected trade secret holder has property-like traits, as far as the allocation of economic value is concerned. Although according to the prevailing opinion no transferable right results, know-how is at least factually transferable and thus can also be the object of legal transactions and be economically utilised. Accordingly, an allocation of economic value, i.e. the possibility of undue enrichment in the case of injury, is affirmed. When it comes to information which is detachable from the business, such as data from automated measurements, this leads to a *de facto* tradeable legal position.

2.1.3. Difficult Determination of the Protected Person

The protection of trade secrets does not cause an independent legal allocation; it merely amplifies an existing *de facto* exclusivity of data by protective rights. So one can indeed refer to the protection of corporate secrets as legal assignment to those parties who have *de facto* exclusive access to the data (by recording or storing it). However, any problems with the existing factual situation tend to be reinforced thereby.

With Industrie 4.0 scenarios the allocation by the protection of trade secrets leads to problems: Usually, there are several businesses involved, whose respective secrets are difficult to distinguish. From the point of view of the user of complex manufacturing machines any generated data are the user's trade secrets. From the point of view of the manufacturer(s) of the machines, that may by design exclude the users from access, it is rather assumed to be their secrets. If several machines work together or are linked, which is likely to be the rule, the allocation becomes even more difficult.

The spatial expansion of the business sphere, i.e. the impossibility to demarcate it by physical gates is an inevitable consequence of IT (see cloud computing). What actually creates problems is the delimitation of one operating sphere to the other. A lack of transparency is one of the main problems with modern IT issues. This is true not only for consumers but also for companies. And this is exacerbated by the protection of corporate secrets, which argue for a cautious application of trade secret protection to Big Data matters.

2.1.4. Complex Contracting

Trade secrets generally require more complex contracts than licences relying on IP. This may favour contracting parties who can afford professional legal expertise. However, one may argue that this is no problem for business to

---

[31] Council Regulation (EC) No 6/2002 of 12.12.2001 on Community designs [2002] OJ L3/1.

business contracts whereas with business to consumer contracts trade secret protection does not apply for the consumer but the consumer being a natural person can rely on data protection.

2.2. TECHNICAL PROTECTION MEASURES: FACTUAL EXCLUSIVITY AND LEGAL PROTECTION

Another possibility to keep data exclusive is the use of technical protection measures. This may allow a contracting party to supply data but keep factual control over the use of the data. There is also an additional legal protection but only in respect of works or other protected subject matter of copyright law.

2.2.1. Factual Exclusivity without Secrecy

Technical protection measures may effect factual exclusivity of digital goods. It may not only be used with physical data carriers but even with any kind of syntactic information. Therefore, technical protection measures have become an important tool for controlling digital goods, be they supplied as physical carriers, downloads or streaming uses.

2.2.2. Legal Protection

The additional legal protection according to Articles et 6 seqq. Infosoc Directive[32] may be regarded as a *de facto* IP right.[33] Its function, however, is closely related to trade secret protection. Factual exclusivity is reinforced by legal prohibition ('adequate legal protection against the circumvention of any effective technological measures', Art. 6 (1)).

With respect to data the application of these rules is limited. Only 'acts, in respect of works or other subject matter, which are not authorised by the holder of any copyright or any right related to copyright as provided for by law or the sui generis right' (Art. 6(3)) are covered. Therefore, databases may enjoy legal protection but not single data (see section 2.4).

2.2.3. Contracts Concerning Protected Data

Contracts tend to be simplified by the application of technical protection measures. In effect digital goods are rendered factually exclusive like physical goods. However, there is a certain danger that technical protection measures are

---

[32] Directive 2001/29/EC on the harmonisation of certain aspects of copyright and related rights in the information society [2001] OJ L167/10.
[33] S. BECHTOLD, *Vom Urheber- zum Informationsrecht*, C.H. Beck, Munich 2001, p. 269 et seqq.

used to unilaterally create IP-like exclusivity without the necessary checks and balances.

Contract law (especially mandatory implied warranties and the control of terms and conditions) may also provide effective control mechanisms.[34] Article 6(1)(a) of the proposed Directive concerning contracts for the supply of digital content[35] may be interpreted in an objective way relying on reasonable customer expectations and used as a tool against the unjustified use of technical protection measures.

2.3. PERSONAL DATA: DATA PROTECTION

Data protection legislation has become the subject of European law and is now fully harmonised by the General Data Protection Regulation.[36] The regulatory purpose of data protection law, as is clarified in the first recital of the Regulation, is the protection of individuals with regard to the processing of personal data. The formulation of the relevant fundamental rights guaranteed in Article 8(1) EU Charter of Fundamental Rights and Article 16(1) TFEU is misleading as regards the wording: 'Everyone has the right to the protection of personal data concerning him or her.' This however does not in particular mean the protection of data, but rather the protection of the person from any danger caused by the use of their personal data.

However, it is also argued that the current data protection law could be further developed into a data property right, meaning that its function is thus expanded from a mere protection of personality to a participation in the economic value of personal data.[37]

When further developing data protection into rights to use data of the persons concerned (data subjects), several problems arise. On the one hand, data protection law does not allocate all personal data exclusively to the person concerned. Data protection law protects the personality only against

---

[34] MACKENRODT, *supra* n. 23.
[35] Proposal for a Directive of the European Parliament and of the Council on certain aspects concerning contracts for the supply of digital content, COM(2015) 634 final.
[36] See *supra* n. 19.
[37] B. BUCHNER, *Informationelle Selbstbestimmung im Privatrecht*, Mohr Siebeck, Tübingen 2006, p. 202 et seqq.; W. KILIAN, 'Wie der Staat unsere Daten schützen kann', *Frankfurter Allgemeine Zeitung* of 04.07.2014; R. SCHWARTMANN and C. HENTSCH, 'Eigentum an Daten – Das Urheberrecht als Pate für ein Datenverwertungsrecht' (2015) *Zeitschrift für Datenschutz-, Informations- und Kommunikationsrecht* 221. For the US discussion regarding data protection as 'quasi IP' see L.H. SCHOLZ, 'Privacy as Quasi-Property' 2015 (2016) *Iowa Law Review*; S. BALGANESH, 'Quasi-Property: Like, But Not Quite Property' (2012) 160 *University of Pennsylvania Law Review* 1889; P. SAMUELSON, 'Privacy as Intellectual Property' (1999) 52 *Stanford Law Review* 1125; Cf. K.v. LEWINSKI, *Die Matrix des Datenschutzes*, Mohr Siebeck, Tübingen 2014, p. 50 et seqq.

certain forms of data processing. On the side of legal consequences of data protection, an allocation of economic value may be affirmed in the way it is affirmed for other personal rights. Despite the lack of transferability, the legal system tolerates the economic utilisation of personal rights, in particular through contracts regulated by the law of obligations.[38] This raises the central question of whether a mere reference to personality actually causes a sufficient justification for the allocation of transferable exclusive rights in dealing with one's own personal data.

2.3.1. No Comprehensive Allocation of Data to the Person Concerned

Data protection law does not allocate comprehensive rights of use relating to one's personal information, but concerns only certain forms of data processing which, based on an overall consideration, entail a serious interference with personal rights. In the relationship between private persons the regulations in Article 6(1)(f) General Data Protection Regulation play a pivotal role. They show that data protection, like many personal rights, is a law open to consideration (i.e. interventions can be justified by balancing against conflicting interests).

A complete allocation of one's own data (semantically defined by reference to their own personality) would be inconsistent with the constitutionally protected freedom of expression and information. The German Federal Supreme Court summarises this succinctly in the *Spickmich.de* judgment: 'However, the individual has no absolute unrestricted control over its data because it develops its personality within the social community. Within a social community information, even if it is personal, is a part of social reality which cannot be exclusively assigned to the person concerned.'[39]

2.3.2. Property-like Allocation of Personal Data?

In terms of legal consequences, the question arises as to whether to assign to the person concerned a property right (allocating the economic value of data to the right-holder). The position procured by current data protection law is certainly

---

[38] A. OHLY, 'Volenti non fit inuria', *Die Einwilligung im Privatrecht*, Mohr Siebeck, Tübingen 2002, p. 141 et seqq., p. 165 et seqq., p. 259 et seqq.; H. BEVERLEY-SMITH, A. OHLY and A. LUCAS-SCHLOETTER, *Privacy, Property and Personality*, Cambridge University Press, Cambridge 2005, p. 94 et seqq.; F. HOFMANN, 'The economic part of the right to personality as an intellectual property right? A comparison between English and German Law' (2010) 2 *ZGE/IPJ* 1.

[39] BGHZ 23.06.2006 181, 328 – spickmich.de. 'Allerdings hat der Einzelne keine absolute, uneingeschränkte Herrschaft über 'seine' Daten; denn er entfaltet seine Persönlichkeit innerhalb der sozialen Gemeinschaft. In dieser stellt die Information, auch soweit sie personenbezogen ist, einen Teil der sozialen Realität dar, der nicht ausschließlich dem Betroffenen allein zugeordnet werden kann.'

not transferable (unlike the right of publicity recognised in the US). However, since with the consent of the person concerned illegal data processing becomes legal according to Article 6(1)(a) General Data Protection Regulation, there is the possibility to grant such consent and by doing so enter into a contractual obligation within the limits of contract control.[40] Thus the concerned party can use its position similar to other personal rights ('personality licensing').

However, there is definitely a 'need to develop general rules on personal data as an object of contractual obligations and as consideration.'[41] In particular, the legislator needs to address the question of to what extent data protection may be regarded as a property-like right. Under the current law contracting is limited with good cause: the protection of the person concerned. The possibility to conclude binding agreements is severely limited by the free revocability of consent which is also codified in Article 7(3)(1) General Data Protection Regulation:

> 'The data subject shall have the right to withdraw his or her consent at any time. The withdrawal of consent shall not affect the lawfulness of processing based on consent before its withdrawal. Prior to giving consent, the data subject shall be informed thereof. It shall be as easy to withdraw as to give consent.'

Despite the option to revoke the consent, the general possibility to exploit protected personal data by granting consent remains,[42] especially since the revocation is not retroactive.

A second restriction may arise if consent according to Article 7(4) General Data Protection Regulation only contains an unclear compromise:

> 'When assessing whether consent is freely given, utmost account shall be taken of whether, *inter alia*, the performance of a contract, including the provision of a service, is conditional on consent to the processing of personal data that is not necessary for the performance of this contract.'[43]

This leaves open the question of how the rule will be handled in practice. It does not preclude any contracts where consent is given as consideration. But it does not give clear guidelines when such consent shall be regarded as freely given and when not.

---

[40] LANGHANKE and SCHMIDT-KESSEL, *supra* n. 22, p. 218, 220: 'Therefore, not the supply of the data as such is usually of much interest to the creditor. Rather, he is interested in becoming justified for certain kinds of processing the data.'
[41] *Ibid.*, p. 218.
[42] Cf. *ibid.*, p. 221.
[43] Article 7(4) of a leaked older version read: 'When assessing whether consent is freely given, utmost account shall be taken of whether, among others, the performance of a contract, including the provision of a service, is made conditional on the consent to the processing of data that is not necessary for the performance of that contract.'

The possibility to enter into contracts concerning data does not make data protection a property right. There is still the balancing of interests resulting from data protection law. So one can at best speak of a framework 'right to the protection of personal data' (i.e. open to balancing with conflicting interests) but not of a property-like 'right to one's own datum'.[44] This is in the best interest of the persons concerned. Unlike with property rights nobody can be forced into losing her or his protection.

2.3.3. Is Data Protection as a Property Justified?

In the background of the discussion about IP-like effects of data protection is an issue of justification. Is the mere fact that an economic good has references to a person a reasonable justification for the participation its economic value? Copyright, which is often mentioned as a paradigm for personality-related property (according to the ruling monistic theory in Germany), has a different function. On the one hand, the required personality reference for copyrighted works is sometimes very limited. On the other hand, the creation of a new good is an additional crucial aspect for the justification of copyright, which also allows copyright to be justified by economic reasoning.

The strongest argument may be that real property-like rights can be withdrawn even against the will of the entitled legal entity, for example in the event of foreclosures.[45] This must not be possible with personal rights, at least if the legal entity itself has not utilised them (similar to copyright). With data protection, where the excessive grant of consents by means of forms is often classified as 'consideration' for services free of charge, with regard to the social customary and often lack of alternatives, such consent already appears problematic.

2.4. DATABASES: RIGHT FOR THE MAKER OF A DATABASE

With the *sui generis* protection for the maker of a database according to Article 7 et seqq. of Directive 96/9/EC,[46] the European legislator already

---

[44] e.g. K.-H. LADEUR, 'Datenschutz – vom Abwehrrecht zur planerischen Optimierung von Wissensnetzwerken' (2000) *Datenschutz und Datensicherheit* 12, 18: '*nicht dem Persönlichkeitsrecht zugeordnet ..., sondern als Bestandteil eines neuartigen Eigentumsrechts*' ('not a personal right but part of a new kind of property right').

[45] A. PEUKERT, *Güterzuordnung als Rechtsprinzip*, Mohr Siebeck, Tübingen 2008, p. 534 et seqq.

[46] Directive 1996/9/EC of the European Parliament and of the Council of 11 March 1996 on the legal protection of databases [1996] OJ L77/20. This *sui generis* right supplements the already existing copyright protection of databases (§4 II UrhG/German Copyright Act). The requirement of a substantial investment (§87a I UrhG/German Copyright Act) replaces the requirements for copyright protection of a personal intellectual creation ('*persönliche geistige Schöpfung*') involving a selection or an arrangement. The exclusive right is assigned to the investor (maker of a database, §87a II UrhG/German Copyright Act).

addressed the problem of legally capturing data as an economic good. The subject matter of the right, however, is not the data themselves, but the database or the investment in the creation of a database. Unlike traditional intellectual property rights this law no longer protects a creation of the human mind, but rather the result of an investment.[47] Case law has shaped the right for the maker of a database in a way that it generally does not encompass the use of data in Big Data issues, at least not for the data generator.[48] This is mainly due to the definition of significant investment as a condition of protection which is reflected in the scope of protection.

2.4.1. Collections of Data as Databases

With automatically collected data no human intellectual achievement is present, which is why in accordance with Article 3(1) Directive 96/9/EC no database work is protected by copyright. By contrast, the outcome of automated measurements and recordings saved to a database usually qualifies as a database (Article 1(2) Directive 96/9/EC).

---

[47] A. STEINBECK, 'Immaterialgüterrechte und Informationsinteresse' (2010) *Kölner Schrift zum Wirtschaftsrecht* 223, 224: '*Nicht nur der Schöpfer einer geistigen Leistung, sondern auch derjenige, der mit organisatorischem Einsatz und finanziellen Investitionen dazu beiträgt, dass Informationen generiert und/oder verbreitet werden, hat regelmäßig ein Interesse daran, diese Informationen zunächst ausschließlich selbst zu nutzen und auszuwerten*' ('Not only the creator of an intellectual effort but also everyone supporting the creation and/or dissemination of information by contributing organisational effort and financial investment possesses a general interest in initially using and exploiting this information on an exclusive basis').

[48] C. ZIEGER and N. SMIRRA, 'Fallstricke bei Big Data-Anwendungen – Rechtliche Gesichtspunkte bei der Analyse fremder Datenbestände' (2013) *MMR (Multimedia und Recht Zeitschrift für Informations-, Telekommunikations- und Medienrecht)* 418, 420: '*relevante Verletzungen*' als '*Ausnahme*' ('relevant violation' as an 'exception'); A. WIEBE, *supra* n. 16, p. 1 et seq., however, shows that there are many Big Data situations without automated measurements and that the *sui generis* right is highly relevant for Big Data situations (although not for the protection of the data producer). Similar T. EHMANN, 'Big Data auf unsicherer Grundlage – was ist "wesentlich" beim Investitionsschutz für Datenbanken?' (2014) 6 *K&R (Kommunikation & Recht)* 394, 395: '*Die ... ergangenen höchstrichterlichen Entscheidungen haben dem neuen Schutzrecht inzwischen erste Konturen verliehen. Dennoch hinkt die Bedeutung des Datenbankherstellerrechts in der Rechtspraxis der wirtschaftlichen Bedeutung von Daten als Wirtschaftsgut hinterher*' ('The ... case law has begun to shape the new protective right. Nevertheless, the impact of the database producer right on the legal practice still lags behind the economic significance of data as an asset'). It could be argued that the importance in legal practice lags behind precisely because of the contouring, in particular due to Case C-203/02, *The British Horseracing Board Ltd and Others v. William Hill Organization Ltd*, ECLI:EU:C:2004:695.

2.4.1.1. Collection

Especially with automated measurements, a useless collection of data is not what is produced, but rather a structured arrangement of data.[49] The files to be transferred, containing the data of individual devices and *a fortiori* files created by merging data from multiple devices, should in any case be qualified as a collection.

2.4.1.2. Of Independent Works, Data or Other Materials

Independent materials of this collection are the data as single measurements, i.e. differentiated on the semantic level. The legislator's decision not to base the subject matter on individual data is likely to be due to the fact that the legislator wanted to keep the (semantic) information in the public domain.[50] An exclusive right to the individual semantic information (as was discussed, for example, for the protection of daily updated news)[51] cannot be justified. It would have been easier to base protection on the individual data and to distinguish them on the syntactic level.

The independence of the materials which is required to qualify the collection as a database is not a problem; it is almost a hallmark of Big Data cases where a large amount of data is not intended for a specific use at first but is optionally analysed later. Based on the *Fixtures Marketing* jurisprudence,[52] independence presupposes that the individual elements can be used on their own. This is true for even the most insignificant datum within the meaning of a statement, to the extent that it is a well-structured information unit,[53] i.e. a unit with semantic content.[54] With the formulation 'independent works, data or other materials' (German version: *'Daten oder andere unabhängige Elemente'*, data or other independent elements) the law assumes that at least semantically determined data are automatically independent.

---

[49] A. WIEBE, *supra* n. 16, p. 2.
[50] D. THUM and K. HERMES in A.-A. WANDTKE and W. BULLINGER (eds.), *UrhR*, 4th ed., C.H. Beck, Munich 2014, §87a marginal no. 26; EHMANN, *supra* n. 48, p. 395: no 'monopolisation of (important) information' (*keine 'Monopolisierung von (wichtigen) Informationen'*). A. WIEBE in G. SPINDLER and F. SCHUSTER (eds.), *Recht der elektronischen Medien*, 3rd ed., C.H. Beck, Munich 2015, §87a marginal no. 1.
[51] See also H. PRANTL, *Die journalistische Information zwischen Ausschlussrecht und Gemeinfreiheit*, Gieseking, Bielefeld 1983, p. 36 et seqq.
[52] Case C-338/02, *Fixtures Marketing Ltd v. Svenska Spel AB*, ECLI:EU:C:2004:696; Case C-444/02, *Fixtures Marketing Ltd v. Organismos prognostikon agonon podosfairou AE (OPAP)*, ECLI:EU:C:2004:697; Case C-46/02, *Fixtures Marketing Ltd v. Oy Veikkaus Ab*, ECLI:EU:C:2004:694.
[53] THUM and HERMES, *supra* n. 50, §87a marginal no. 13.
[54] To that effect also T. DREIER in T. DREIER and G. SCHULZE (eds.), *UrhG*, 5th ed., C.H. Beck, Munich 2015, §87a marginal no. 6.

With the criterion of independence understood as independent usability, the concept of goods comes into play – in other words not only the database, but also the single datum or element, should represent independent goods. This demonstrates the classic understanding of data use on which the regulation is based: collecting and processing go hand in hand and as a result provide a collection of data easily perceivable by users which are therefore usable data. The Big Data paradigm, however, is the exact opposite: even the most insignificant raw data represent goods that can only be used by subsequent analysis.[55]

2.4.1.3. Systematic or Methodical Arrangement

In order to use individual data it is vital that databases are structured systematically or methodically as part of the data collection. This minor problem is solved if the individual datum is filed in a way that it can be retrieved.[56] Thus, to meet this criterion, data have to be compiled in a way that preserves the semantic value of the data. It is, however, extremely unlikely for a collection of data to be arranged in such a way that the entire collection can no longer be used, and thus the requirement of a systematic or methodical arrangement becomes a *de minimis* criterion.[57]

2.4.2. Protection Requirement of Substantial Investment

As the investment represents a protection requirement, the *sui generis* protection is quite similar to unfair competition law; however, it is not as flexible.[58] The main reason why *sui generis* protection has not had any significant impact in Big Data applications might be that the ECJ in its decisions *BHB/Hill* and *Fixtures Marketing I–III* established a limitation on investments in existing data.[59] Thus, data producers cannot be a legal entity. Although it narrows down the wide potential application of database protection, that the reasons why investments in data production are excluded are hard to understand.[60] The production of data

---

[55] NEUMANN, *supra* n. 27.
[56] DREIER, *supra* n. 54, §87a marginal no. 7; THUM and HERMES, *supra* n. 50, §87a marginal no. 21; A. AUER-REINSDORFF, 'Schutz in Datenbanken und Datenbankwerken' in I. CONRAD and M. GRÜTZMACHER, *Recht der Daten und Datenbanken im Unternehmen*, Otto Schmidt, Cologne 2014.
[57] THUM and HERMES, *supra* n. 50, §87a marginal no. 24.
[58] EHMANN, *supra* n. 48, p. 399; cf. M. LEISTNER in O. TEPLITZKY and K.-N. PEIFER and M. LEISTNER (eds.), *GK-UWG*, 2nd ed., De Gruyter, Berlin 2013, §4 no. 9 marginal no. 84, p. 131.
[59] *Supra* n. 52; see also M. LEISTNER, 'Urteilsanmerkung zu EuGH BHB v William Hill' (2005) 60 *Juristenzeitung* 408, 409; M. LEHMANN, 'Abgrenzung der Schutzgüter im Zusammenhang mit Daten' in I. CONRAD and M. GRÜTZMACHER, *supra* n. 56, p. 133, 138 et seqq.; EHMANN, *supra* n. 48, p. 397 et seq.; A. WIEBE, *supra* n. 16, p. 4.
[60] T. HOEREN 'Comment on Case C-203/02, The British Horseracing Board Ltd. u.a. v. William Hill Organization Ltd' (2005) *MMR* 34, 35; M. LEISTNER, 'The protection of databases' in

could have easily been interpreted as acquisition. However, the main function of this restrictive interpretation is to avoid an exclusivity right in information which cannot be obtained otherwise. This may well hinder the protection of measurement data when the process or object under observation is no longer available. The German Federal Court, however, ruled that the detection of factual processes only concerns pre-existing information which consequently is not produced.[61] As a consequence music hit lists may be protected due to the investment in ascertaining sales figures and radio playlists.

As for the question whether side products of entrepreneurial activity ('spin-offs') can be the subject of necessary investment, another problem arises.[62] An important characteristic of Industrie 4.0 matters is that the data are created incidentally. In its *Autobahnmaut* (motorway charge) decision[63] the German Federal Court of Justice had to judge a matter in which the database could only be qualified as a side product and affirmed a substantial investment in the acquisition of data.

Irrespective of whether side products of entrepreneurial activity are subject to the regulations, the criterion of substantial investment excludes consumers from protection.[64] In the context of Big Data, this is problematic. As illustrated below, the protection for data producers could also be outlined without considering investments (be it in production or in the collection of data), thus including consumers as well.

2.4.3. No Exclusive Right in Individual Data

The rights conferred only pertain to the object of protection, i.e. the data collection. Individual elements are not protected. As a precondition for infringement, a significant part of the data, either in terms of quantity (volume of data) or quality (volume of investment), has to be transferred (extraction, Article 7(2)(a) Directive 96/9/EC). However, in such cases the alternative of repeated and systematic uses which conflict with a normal analysis of the database or unreasonably prejudice the legitimate interests of the maker of the

---

E. DERCLAYE, *Research Handbook on the Future of EU Copyright*, Edward Elgar, Cheltenham 2009, p. 427, 438; EHMANN, *supra* n. 48, p. 397: 'Nach dem natürlichen Wortsinn ist die Neugewinnung von Daten ebenso eine Form des Beschaffens wie der Erwerb von Daten' ('According to the natural literal sense gaining new data is a form of procurement as is the purchase of data').

61 BGH, 21.07.2005 – I ZR 290/02 (OLG München) *HIT BILANZ*, GRUR 2005, 857, 858 et seqq.
62 Cf. DREIER, *supra* n. 54, §87a marginal no. 13; EHMANN, *supra* n. 48, p. 397 et seq.
63 BGH, 25.03.2010 – I ZR 47/08 (OLG Hamburg) *Autobahnmaut*, GRUR 2010, 1004.
64 However, if a benefit is granted for the production of data (use of a data collecting equipment), it would be conceivable to assume an investment of consumers. This investment would have to reach such an extent that an independent economic good results (cf. A. AUER-REINSDORFF, *supra* n. 56, p. 205, 215 et seq.), which also applies to raw data.

database can ensure legal protection.[65] For an extraction it is sufficient that the sum of transfers remains below the materiality threshold but represents a 'repeated and systematic extraction and/or re-utilisation of insubstantial parts of the contents of the database implying acts which conflict with a normal exploitation of that database or which unreasonably prejudice the legitimate interests of the maker of the database' (Article 7(5) Directive 96/9/EC). It suffices if the ongoing transfers are aimed at the transfer of essential parts and in the case of its continuation would result in the fact that a substantial part of the database would be extracted.[66]

3. INTRODUCTION OF A DATA PRODUCER RIGHT?

3.1. FEATURES OF A DATA PRODUCER RIGHT

After analysing the existing legal situation, the question arises whether a real right to use data should be created. What such a data producer right (or data producer's right) might look like shall be outlined as briefly as possible below.[67] It should deal with subject matter, conditions, ownership, scope of protection, exceptions, transferability, the relationship with other legal regimes (like data protection), and last but not least the question of justification.

3.1.1. Subject Matter

A well-defined subject matter would be machine-readable coded information that is defined only by its representative characters (bits) irrespective of its content (data delimited on the syntactic level).

3.1.2. Protection Requirements

An important protection requirement would be the creation of the data, meaning creation through automated measurement processes, intellectual activity or simple computing power. An additional possibility would be a limitation on measurement processes to completely exclude intellectual creations and separately regulate the problem of digital goods which are produced by processing power (particularly bitcoins). Comparable current regulations are the protection of photographers under German law (§72 German Copyright Act)

---

[65] A. AUER-REINSDORFF, *supra* n. 56, p. 205, 219.
[66] BGH, 01.12.2010 – I ZR 196/08 (OLG Köln) *Zweite Zahnarztmeinung II*, *GRUR* 2011, 724, 726 et seq.
[67] See also H. ZECH, 'Daten als Wirtschaftsgut – Überlegungen zu einem "Recht des Datenerzeugers"' (2015) *CR* 137, 144 et seqq.

and the protection of phonogram producers (§85 German Copyright Act, cf. Art. 2(d) WIPO Performances and Phonograms Treaty).

3.1.3. First Ownership (Original Rights Holder)

The ownership of the right would be tied to the economically responsible operator of equipment that generates the data (data producer).[68] When determining the rights holder by ascertaining the economic responsibility for the processing of goods one can find parallels in the determination of the processor according to §950 German Civil Code (acquiring property by processing tangible goods) or the person storing data protected under criminal law pursuant to §202a and §303a German Criminal Code. No distinction should be made between data produced by entrepreneurial activities and by consumer behaviour.[69]

3.1.4. Scope of Protection, Exceptions and Limitations

The scope of protection would in particular include use by carrying out statistical analyses, but not the re-creation of the same data by independent measurement. Here there is a parallel to the already mentioned Article 19(2) Regulation 6/2002 on Community designs.

---

[68] Similar Maximilian BECKER, personal communication, who points out, that this also leads to a run in parallel with the protection of trade secrets of the machine operator, who 'usually plays the key role in producing the data' ('*idR den größten Anteil an der Datenerzeugung hat*').

[69] Similar T. HOEREN, 'Dateneigentum – Versuch einer Anwendung von §303a StGB im Zivilrecht' (2013) *MMR* 2013, 486, 487; G. HORNUNG and T. GOEBLE (2015) *CR* 265, 271. Different Maximilian BECKER, personal communication, who instead recommends a 'right to products without data collection' ('*Recht auf datenerhebungsfreie Produkte*') for consumers and a right to use data for businesses only. Argument: 'It argues against an allocation decision, if the person entitled in principle has no use for the allocated good, and it is thus allocated only for the purpose of disposal' ('*Es spricht gegen eine Zuweisungsentscheidung, wenn der Berechtigte prinzipiell keine Verwendung für das zugewiesene Lebensgut hat, es also nur zum Zwecke der Veräußerung zugewiesen wird*'). However, this is also true for example for patents that are granted a construction office; G. HORNUNG and T. GOEBLE (2015) *CR* 265, 272, see political problems: '*So könnte der Gesetzgeber beispielsweise versucht sein, die traditionellen (deutschen) Kfz-Hersteller dadurch vor einem drohenden Bedeutungsverlust zu schützen, dass er ihnen die exklusive Nutzungsbefugnis an anonymisierten fahrzeugbezogenen Daten und das Recht zuweist, diese auf den entstehenden Datenmarktplätzen an Dritte zu veräußern. Auf den ersten Blick würde eine solche exklusive Zuweisung die Position der Hersteller stärken und ihnen eine gute Verhandlungsposition für die Preisbildung geben. Bei näherem Hinsehen ist dies jedoch weniger sicher, wenn es um Verhandlungen mit weltweit operierenden Interessenten geht*' ('The legislator could, for example, be trying to protect traditional (German) vehicle manufacturers of an impending loss of importance that it assigns them the exclusive right of use of anonymised vehicle-related data and the right to sell them on the uprising data marketplaces to third parties. At first sight, such an exclusive assignment would strengthen the position of producers and give them a good position to negotiate pricing. On closer inspection, however, this is less sure when it comes to negotiations with worldwide operating interested parties').

Regarding limitations and exceptions, special attention should be paid to the interest of the public domain. A short term of protection seems to be appropriate since using data by analysing them can be done relatively quickly. There is also no need to exclude private parties as possible rights holders but it seems reasonable to grant protection only against commercial infringements. Merely allowing for private use as an exemption (as with database producer rights under §87 c I No. 1 German Copyright Act; Art. 9(a) Directive 96/9/EC: extraction for private purposes) seems too narrow. A possible wording would be for example that in §11 No. 1 of the German Patent Act or §40 No. 1 German Design Act (also Art. 27(a) of the Agreement on a Unified Patent Court: acts done privately and for non-commercial purposes). The cumulative requirement of a 'private domain' should possibly be adapted to the world of data, which can no longer be spatially delimited. An exemption guaranteeing scientific freedom comparable to Article 9(b) Directive 96/9/EC seems to be appropriate.

3.1.5. Transferability

The right should be designed to be transferable, since the creation of markets for data is one of the main purposes of the new law. Another main purpose (probably the decisive one) is the fair and efficient allocation of the benefits of data use. Within the framework of contract law it will be necessary to consider whether allocation to the producer should be a model and accordingly whether unrequited transfer as part of the terms and conditions should be restricted. A corresponding consideration will have to be recognised in an offer of better or cheaper services.[70]

3.1.6. Interaction with Existing Rights

Finally, questions of the interplay with existing rights, particularly data privacy law, will have to be clarified. Here, a juxtaposition of various allocation systems must be assumed. In any case, the right of the data producer cannot displace the protection of the person concerned.

3.2. POSSIBLE JUSTIFICATION: CREATION OF A MARKET FOR DATA (FAIRNESS ASPECTS)

This finally leads to the question whether such a right should be created or not. On the one hand, there are significant concerns about the resulting restrictions

---

[70] Possible contractual models: pricing with two different prices (one without and one with transfer of future rights to the produced data); lease contract (economic operator is the lessee); or rental contract (economic operator is the landlord).

of the public domain which will be addressed in section 3.3. On the other hand, there are several arguments in favour of a property right in data.

3.2.1. Incentives for Generating and Disclosing Data

As in traditional intellectual property rights, the argument can be made that incentives are created to generate and to reveal data (and hence, indirectly, to promote innovations that are made possible through the use of data) and that markets are created for information goods (that otherwise would not be tradeable or would only be tradeable with higher transaction costs). This function may be less important when it comes to data since data are not intellectual creations, and with technological developments can be produced cheaper and cheaper. In many scenarios it might be argued that the data are produced anyway. In other scenarios where measuring devices are still expensive and costly to run (like a computed tomography scanner or a satellite scanner) the production of data requires huge investments. The combination of necessary investments and easy copying might lead to potential market failures and in turn make legal protection necessary.

3.2.2. Creating a Market for Raw Data in Big Data Scenarios

With the widespread use of Big Data applications, data are also traded as novel goods of their own kind. A data economy not only means a data-driven or data-controlled economy (the control of economic processes through data has existed for a long time and by networking has improved in quality), but also an economy with data as goods. Property rights may lower the transaction costs.

The role of law in a data market can be looked at from two different points of view, corresponding to certain functions of the law. On the one hand, the law can be understood as a restriction on free data traffic. Because of the EU's primary objective of eliminating trade barriers, this is paramount for the DSMS.[71] Such restrictions arise from regulatory law, in particular data protection law, but also as a result of intellectual property rights. On the other hand, it is the law that enables data traffic in the first place. This applies to regulatory law through creation of clear 'rules of the game', but especially to contract law, competition law and intellectual property rights, which create markets for incorporeal assets by allocating transferable rights. Thus the main focus is not to keep the legal framework for a data economy as slim as possible, but to create the appropriate

---

[71] European legislation concerning exclusive rights usually refers to Article 114 TFEU as the basis of competence. Also data protection aims at eliminating trade barriers. The DSMS (p. 14) argues from this angle too, but also states that there is a 'lack of clarity over rights to use data' which not only concerns their function as a market barrier but also as an instrument for the creation of markets.

legal framework for a data economy and in particular for a functioning data market.

The key question about market regulation is whether data without property can be traded or not. Secrecy and technical protection measures might be sufficient to allow contracts for the supply of data to be entered into. The easy copying of data might be a hindrance. Whether functioning markets for raw data require legal protection has to be researched empirically. Most likely, the picture will be different depending on the specific kind of data in question.

3.2.3. Clear Allocation of First Ownership

Another important aspect seems to be that such legal regulation would clearly determine who benefits from the use of data. This would prevent machines being designed in such a way that the data are difficult to read, or other mechanisms being created that grant *de facto* exclusivity (leading to an arms race). Such a regulation may not only save costs, but would also promote a culture of transparency, as 'open data' does. The data producer right would have the same function for 'open data' as copyright has for 'open source' and 'open content'.

3.3. CONCERNS: FREE FLOW OF DATA

The main concern with data-related property rights is that it might have a detrimental effect on the use of existing data. The OECD study 'Data-Driven Innovation – Big Data for Growth and Well-Being', arguably the most influential study on data, sees 'data ownership' as problematic.[72] It states that, due to the many potential rights holders, 'property rights may not be the optimal solution in the case of data'. The OECD sees the 'free flow of data' as a major a goal.[73] Rights inhibiting this flow are seen as a risk. This argument is bolstered by the fact that any new right will act cumulatively with existing rights like data protection. In the DSMS a 'Free flow of data' initiative has been announced. It is to deal with restrictions on the free movement of data for reasons other than the protection of personal data and unjustified restrictions regarding the physical location of storage and processing of data, explicitly addressing *inter alia* the 'emerging issues of ownership'.[74]

It needs to be established empirically whether the existing legal framework is sufficient. If not, the areas where new legal instruments are required need to be precisely defined. Amending other areas of law than property rights may suffice;

---

[72] OECD, *Data-Driven Innovation – Big Data for Growth and Well-Being*, OECD Publishing, Paris 2015, p. 195 et seqq.
[73] *Ibid.*, p. 35.
[74] COM(2015) 192 final, p. 15.

For example, trade secrets protection could suffice in business-to-business contexts, and data protection in business-to-consumer context; contracts can resolve problems of precisely assigning specific uses and rights to interested parties; and contract law could ensure fairness and consumer protection. The situation is still evolving, so it might be too early for legal responses.

4. CONCLUSION

Legal rules governing data contracts are still in their infancy. Before a 'a specific contract law for data' is conceivable the legal status of data should be clarified. *De lege lata* no data property right exists. *De lege ferenda* a data producer right might help create a market for raw data but also meets serious concerns. The development of clear legal rules concerning data as a commodity is a necessity. How such rules might be designed, whether they should include a data property right or not, who should be the original owner of such property and what contract rules concerning fairness and limitations have to be implemented has to be openly debated and decided on a legislative level.

JURISDICTION REGARDING CLAIMS FOR THE INFRINGEMENT OF PRIVACY RIGHTS UNDER THE GENERAL DATA PROTECTION REGULATION

Pietro Franzina

1. Introductory Remarks 82
   1.1. Data Protection and the Law of Contracts 82
   1.2. Object and Purpose 83
2. Jurisdiction and the Right to an Effective Judicial Remedy 85
   2.1. The 'Instrumentality' of Article 79(2) GDPR 85
   2.2. The Right to an Effective Judicial Remedy in the EU Legal Order .. 85
   2.3. The EU Approach to Civil Jurisdiction 86
3. Jurisdiction under the Brussels I *bis* Regulation 88
   3.1. The Conditions of Applicability of the Uniform Rules on Jurisdiction in the Brussels I *bis* Regulation 88
   3.2. General Jurisdiction at the Domicile of the Defendant 88
   3.3. Special Jurisdiction over Matters Relating to Torts 89
   3.4. Special Jurisdiction over Disputes Arising out of the Operation of an Establishment 93
   3.5. Jurisdiction over Co-Defendants 93
   3.6. Explicit and Tacit Prorogation of Jurisdiction 94
   3.7. Jurisdiction to Order Provisional Measures 95
   3.8. *Lis Pendens* 96
4. Jurisdiction over the Infringement of Privacy Rights Pursuant to Article 79(2) GDPR 96
   4.1. The Scope of Application of Article 79(2) GDPR 96
   4.2. The Protective Policy Underlying Article 79(2) GDPR 97
   4.3. The Jurisdiction of the Courts of the Member State where the Establishment of the Controller is Situated 99
   4.4. Jurisdiction at the Habitual Residence of the Data Subject .. 101
5. The Coordination between Article 79(2) GDPR and the Brussels I *bis* Regulation 103

5.1. Assessing whether the Application of a Provision in the
Brussels I *bis* Regulation would be Prejudicial to the GDPR...... 103
5.2. General Rules that do not Appear to Prejudice Article 79(2)
GDPR 104
5.2. General Rules that Appear to Prejudice Article 79(2) GDPR 105
5.3. General Rules whose Applicability to Proceedings Envisaged
in Article 79(2) GDPR Appears to Be Uncertain................ 105

1. INTRODUCTORY REMARKS

1.1. DATA PROTECTION AND THE LAW OF CONTRACTS

Personal data are often collected upon the conclusion of a contract or in connection with the performance of the obligations arising thereunder. One party may need information regarding the other as a means to assess the legal implications of an envisaged contract, or to discharge its obligations towards the latter once the contract is concluded. Information collected in the context of a contract may equally serve purposes unrelated to the particular transaction in question, such as making predictions as to the behaviour of a group of market actors, based on the aggregation, mining and cleansing of personal data (profiling).

As a matter of EU law, issues relating to data protection fall outside the law of contracts. The liability arising from the unlawful collection or processing of personal information does not become contractual in nature just because the infringement occurred in the course of pre-contractual dealings or in relation to the fulfilment of a contractual promise. Different views have been expressed by scholars as to the distinction between contractual and non-contractual liability within the legal order of the EU.[1] The Court of Justice, however, has made clear that when a person claims the liability of another, the mere fact that the two persons are bound by a contract is not sufficient to characterise the claim as a claim in contract. A contractual characterisation, the Court noted, can in fact be retained 'only where the conduct complained of may be considered a breach of contract, which may be established by taking into account the purpose of the contract', as is notably the case 'where the interpretation of the contract

---

[1] See generally on this topic, M. BUSSANI and V.V. PALMER, 'The Frontier between Contractual and Tortious Liability in Europe: Insights from the Case of Compensation for Pure Economic Loss' in A.S. HARTKAMP et al. (eds.), *Towards a European Civil Code*, 4th ed., Alphen aan den Rijn, Kluwer Law International 2011, p. 945. See also O. MORÉTEAU, 'Revisiting the Grey Zone Between Contract and Tort: the Role of Estoppel and Reliance in Mapping Out the Law of Obligations' in H. KOZIOL and B.C. STEININGER (eds.), *European Tort Law 2004*, Vienna – New York, Springer 2005, p. 60.

[linking one party to the other] is indispensable to establish the lawful or, on the contrary, unlawful nature' of the conduct at stake.[2] The finding, it is submitted, applies irrespective of whether the contract against the background of which the infringement occurred is a business-to-business contract or a consumer contract. Actually, as the Court stated with respect to jurisdiction over cross-border consumer disputes, the EU rules aimed at the protection of consumers' rights only apply to such claims as '*relate* to a contract which has been concluded between a consumer and a professional'.[3] Thus, they only apply when a contract has in fact been entered into by the parties, and the relief sought is based on that contract.

All this notwithstanding, privacy law is relevant to contractual practice in various respects. On the one hand, the way in which personal data is collected and processed affects consumer confidence and may influence the choices of informed and empowered consumers as regards the conclusion of contracts for the supply of goods and the provision of services. On the other hand, the combined consideration of contract rules and the rules concerning the protection of personal data will normally prove crucial to traders (and platform operators, as the case may be) who wish to improve their communication strategies, in particular by fulfilling in a coherent and coordinated way the (pre-)contractual and privacy-related duties that they are required to comply with in their day-to-day business.

1.2. OBJECT AND PURPOSE

This contribution is concerned with the rules that govern adjudicatory jurisdiction in respect of claims based on the alleged infringement of privacy rights, i.e. the rules that determine which court, or courts, are entitled to decide the substance of a cross-border dispute (e.g. where the data subject seeks reparation from a controller whose statutory seat or administration is in a country other than the country where he lives).

Specifically, the focus is on the provision laid down for this purpose in Regulation (EU) 2016/679 of 27 April 2016 on the protection of natural persons with regard to the processing of personal data and on the free movement of such data (the General Data Protection Regulation, or GDPR).[4] According to

---

[2]   Case C-548/12, *Marc Brogsitter v. Fabrication de Montres Normandes EURL, Karsten Fräßdorf*, ECLI:EU:C:2014:148, paras. 23–25. The issue addressed by the Court on that occasion concerned the interpretation of the expression 'matters relating to a contract' as employed in Article 5(1) of Regulation (EC) No. 44/2001 of 22 December 2000 on jurisdiction and the recognition and enforcement of judgments in civil and commercial matters.

[3]   Case C-180/06, *Renate Ilsinger v. Martin Dreschers*, ECLI:EU:C:2009:303, para. 52 (emphasis added).

[4]   [2016] OJ L119.

Article 79(2) GDPR, '[p]roceedings against a controller or a processor shall be brought before the courts of the Member State where the controller or processor has an establishment', or, '[a]lternatively, … before the courts of the Member State where the data subject has his or her habitual residence'.[5] The provision does not apply if 'the controller or processor is a public authority of a Member State acting in the exercise of its public powers'.

Article 79(2) GDPR does not include a complete and self-sufficient jurisdictional regime applicable to the kind of proceedings that come with its scope of application. In these circumstances, the rules that currently apply in the Member States to determine the jurisdiction of courts over the infringement of privacy rights, that is the rules laid down in Regulation (EU) 1215/2012 of 12 December 2012 on jurisdiction and the recognition and enforcement of judgments in civil and commercial matters (the Brussels I *bis* Regulation),[6] are set to maintain, in this area, some of their practical relevance. Clearly, after the GDPR becomes applicable, on 25 May 2018, the Brussels I *bis* Regulation will shift from centre-stage to the background. As stated in Article 67 Brussels I *bis* Regulation (and recalled in recital 147 of the GDPR itself), the general rules on jurisdiction of the former Regulation 'shall not prejudice' the application of the provision specifically introduced for this purpose by the GDPR. In practice, jurisdiction will need to be established pursuant to Article 79(2) GDPR, as complemented by such general rules in the Brussels I *bis* Regulation as do not prejudice the operation of the GDPR.

The present contribution discusses the main questions raised by Article 79(2) GDPR, including the issue of its coordination with the Brussels I *bis* Regulation. The chapter begins by examining the function entrusted with Article 79(2) in the context of the GDPR, and the relevance of the rules regarding jurisdiction to the enjoyment, by data subjects, of the right to an effective judicial remedy against the infringement of the rights enshrined in the GDPR itself. The article goes on to sketch the main rules of jurisdiction laid down in the Brussels I *bis* Regulation, as they currently apply to proceedings concerning the breach of privacy rights. The chapter then moves on to consider the salient features of the 'derogatory' jurisdictional provision introduced by the GDPR as well as some of the questions that courts and interpreters may need to address in connection with the new rule. The final part of the chapter discusses the relationship

---

[5] Although the provision explicitly contemplates proceedings brought against both 'controllers' and 'processors', the rest of the chapter will, for the sake of simplicity, refer to proceedings against the former alone. Actually, the rules on jurisdiction that apply to processors do not differ from those applicable to controllers. Controllers are defined under Article 4(7) GDPR as the natural or legal persons, public authority, agency or other body which, alone or jointly with others, determine the purposes and means of the processing of personal data, while processors, as stated in Article 4(8), merely process personal data on behalf of a controller.

[6] [2012] OJ L351.

between Article 79(2) GDPR and the Brussels I *bis* Regulation and the issues surrounding the concurrent operation of the two texts.

2. JURISDICTION AND THE RIGHT TO AN EFFECTIVE JUDICIAL REMEDY

2.1. THE 'INSTRUMENTALITY' OF ARTICLE 79(2) GDPR

Article 79(2) presents itself as an instrument to the effective protection of the substantive rights enshrined in the GDPR as regards the collection and processing of personal data. The assumption behind Article 79(2) is that, for privacy rights to be practical and effective, data subjects must be given an actual opportunity to seek a judicial remedy each time they consider that their rights under the GDPR have been violated.

In this respect, Article 79(2) may be seen as a development of Article 22 of Directive 95/46/EC of 24 October 1995 on the protection of individuals with regard to the processing of personal data and on the free movement of such data,[7] the predecessor of the GDPR. The latter provision required Member States to 'provide for the right of every person to a judicial remedy for any breach of the rights guaranteed him by the national law applicable to the processing in question'. Article 79(2) reinforces the right to an effective judicial remedy by designating, through a set of regionally uniform criteria, the courts in fact entitled to grant such a remedy.

2.2. THE RIGHT TO AN EFFECTIVE JUDICIAL REMEDY IN THE EU LEGAL ORDER

Article 47(1) of the Charter of Fundamental Rights of the European Union gives everyone 'whose rights and freedoms guaranteed by the law of the Union are violated' the right to an 'effective remedy before a tribunal'. According to Article 8(1) of the Charter and Article 16(1) of the Treaty on the Functioning of the European Union (TFEU), these rights and freedoms include the right to the protection of personal data.

Along these lines, Article 79(1) GDPR provides that, without prejudice to any other available remedy, including the right to lodge a complaint with the competent supervisory authority, 'each data subject shall have the right to an effective judicial remedy where he or she considers that his or her rights under

---

[7] [1995] OJ L281.

[the GDPR] have been infringed as a result of the processing of his or her personal data in non-compliance with [the] Regulation'.

As clarified by the European Court of Human Rights with reference to Article 6 of the European Convention for the Protection of Human Rights and Fundamental Freedoms (the provision that served as a model for Article 47 of the Charter),[8] the right of access to a court includes, *inter alia*, the following rights: the right for the individual concerned to 'have a clear, practical opportunity to challenge an act that is an interference with his rights',[9] the right to obtain a determination of the dispute by a court,[10] and the right to have any (final) judicial decision enforced,[11] including where enforcement is sought of a judgment rendered abroad.[12]

The right of access to a court is not an absolute right. It calls, by its very nature, for regulation by States (under uniform regional rules, as the case may be). The right to a court may be subject to limitations, but these limitations may be considered to be compatible with the Convention (and the Charter) only if they do not restrict the access left to the individual in such a way or to such an extent that the very essence of the right is impaired.[13]

The rules on adjudicatory jurisdiction are part of the rules that regulate the right of access to a court. They realise (and limit) the right to an effective judicial remedy for the determination of civil rights and obligations by identifying the circumstances in which the authorities of a country are allowed, or required, to entertain a case, and, conversely, the cases in which the same authorities are presumed, or prescribed, to dismiss it. In practice, it is for the rules on jurisdiction to determine whether a dispute is connected with a certain country in a way that justifies a decision by the courts of that country, i.e. whether a 'ground' exists for such courts to assert their jurisdiction over the matter.

2.3. THE EU APPROACH TO CIVIL JURISDICTION

Several EU legislative measures deal with adjudicatory jurisdiction. Most of these measures have been enacted on the basis of Article 81 of the TFEU, on judicial cooperation in civil matters (formerly Article 65 of the Treaty Establishing the European Community, as amended by the Treaty of Amsterdam). One of these measures – actually the cornerstone of the law of judicial cooperation in civil matters in Europe – is the Brussels I *bis* Regulation, which was mentioned above.

---

[8] *Explanations Relating to the Charter of Fundamental Rights* [2007] OJ C307, p. 30.
[9] European Court of Human Rights, 4 December 1995, *Bellet v. France* (no. 23805/94), para. 36.
[10] European Court of Human Rights, 1 March 2002, *Kutić v. Croatia* (no. 48778/99), para. 25.
[11] European Court of Human Rights, 19 March 1997, *Hornsby v. Greece* (no. 18357/91), para. 40.
[12] European Court of Human Rights, 1 April 2010, *Vrbica v. Croatia* (no. 32540/05), para. 62.
[13] European Court of Human Rights, 27 August 1991, *Philis v. Greece* (nos. 12750/87, 13780/88 and 14003/88), para. 59.

The rules on jurisdiction in the Brussels I *bis* Regulation are general and paradigmatic. They are general in the sense that they apply, in principle, to any dispute in the field of civil or commercial law, safe for a limited number of excluded matters, which are listed exhaustively in the Regulation itself. They are paradigmatic in the sense that the rules enshrined in the Brussels I *bis* Regulation, formerly included in the Brussels Convention of 27 September 1968 on jurisdiction and the recognition of judgments in civil and commercial matters and in Regulation (EC) 44/2001 of 22 December 2000,[14] have served as a reference for most of the other rules enacted by the EU legislature to deal with matters of civil jurisdiction. Broadly speaking, they can be regarded as illustrative of the general approach of EU law to adjudicatory jurisdiction.

The EU approach to jurisdiction largely builds on two ideas: the idea of abstract legal certainty and the idea of proximity.

Legal certainty requires the court seised of the matter to ascertain its jurisdiction merely on the basis of such predetermined and abstract connecting factors as are set forth by the relevant rules on jurisdiction. Individual courts are granted limited or no discretion to assess, on a case-by-case basis, whether it would be convenient to entertain a claim and decide it on the merits: either a court possesses jurisdiction pursuant to the applicable provisions, and has thus no other option than to exercise it; or it does not possess jurisdiction, in which case it is under an obligation to dismiss the claim.

For its part, the principle of proximity posits that jurisdiction should lie with a court with which the dispute features a close and meaningful connection. As an alternative to jurisdiction being conferred on the courts of the State in which the defendant is domiciled (or where he habitually resides, depending on the subject matter of the dispute), proceedings may be brought, pursuant to the relevant rules of 'special' jurisdiction, before the courts for the place where the disputed relationship is to be localised.

All in all, the combined operation of these two general principles is believed to allow the parties to easily determine the court, or courts, before which the case may be tried, and is considered to provide each litigant with the opportunity to actually take part in proceedings for the determination of their civil rights and obligations without being exposed to unreasonable costs in connection with the internationality of the dispute. Abstract legal certainty combined with proximity is in fact credited with providing equitable and efficient responses to the problems of jurisdiction, while striking a fair balance between the right of the claimant to be granted a practical and effective access to judicial protection, and the right of the defendant not to be subjected to the jurisdictional power of a court that has but a weak, transient or unpredictable connection with the dispute.

---

14 [2001] OJ L12.

3. JURISDICTION UNDER THE BRUSSELS I *BIS* REGULATION

3.1. THE CONDITIONS OF APPLICABILITY OF THE UNIFORM RULES ON JURISDICTION IN THE BRUSSELS I *BIS* REGULATION

The Brussels I *bis* Regulation lays down a comprehensive body of rules dealing with the jurisdiction of Member States' courts as regards proceedings against persons domiciled in a Member State. Actually, as stated in Article 5(1), persons domiciled in a Member State may be sued in the courts of another Member State *only* by virtue of the rules set out in the Regulation itself. Jurisdiction over claims against persons domiciled in a third country (other than a contracting party to the Lugano Convention of 30 October 2007 on jurisdiction, recognition and enforcement of judgments in civil and commercial matters)[15] is asserted, in each Member State, in accordance with the (non-uniform) rules of jurisdiction of the forum. In these instances, the latter rules are understood to apply by virtue of the Regulation, rather than in their own right.[16]

For the purposes of the Regulation, the domicile of a person – namely a company or another legal person – must be deemed to coincide with the place where the company or legal person in question has its statutory seat, its central administration, or its principal place of business.[17] If these places are located in different States, then the person will be regarded as domiciled in each of these States.

3.2. GENERAL JURISDICTION AT THE DOMICILE OF THE DEFENDANT

Under Article 4(1) Brussels I *bis* Regulation, 'persons domiciled in a Member State shall, whatever their nationality, be sued in the courts of that Member State'. The provision confers 'general' jurisdiction. It applies to any claim falling within the scope of the Regulation, with the sole exception of claims for which 'exclusive' jurisdiction is conferred on the courts of a Member State pursuant to Articles 24 and 25.

---

[15] [2009] OJ L147.
[16] Court of Justice, Opinion 1/03 of 7 February 2006, *Competence of the Community to conclude the new Lugano Convention on jurisdiction and the recognition and enforcement of judgments in civil and commercial matters*, ECLI:EU:C:2006:81, para. 148.
[17] Article 63(2) clarifies that, for the purposes of Ireland, Cyprus and the United Kingdom, the expression 'statutory seat' refers to the registered office or, where there is no such office anywhere, the place of incorporation or, where there is no such place anywhere, the place under the law of which the formation took place.

In practice, a data subject who considers that his or her rights have been impaired as a result of the unlawful collection or processing of information may, under the Regulation, sue the data controller before the Member State in which the latter has its statutory seat, its central administration, or its principal place of business.

3.3. SPECIAL JURISDICTION OVER MATTERS RELATING TO TORTS

Article 7 Brussels I *bis* Regulation lists a number of 'special' heads of jurisdiction. These allow the plaintiff to sue a person domiciled in a Member State before the courts of a different Member State, on grounds of proximity. The various heads of jurisdiction contemplated in Article 7 are 'special' in that each of them applies to a given class of claims. The special grounds of jurisdiction are 'alternative' and 'optional', since it is for the plaintiff to decide whether to bring proceedings in the country where the defendant is domiciled, under Article 4(1), or before the courts for the place where the subject matter of the dispute appears to have its centre of gravity, pursuant to Article 7.

As regards 'matters relating to tort, delict or quasi-delict', Article 7(2) confers special jurisdiction to the 'courts for the place where the harmful event occurred or may occur'.

The Court of Justice clarified in *Mines de Potasse d'Alsace* that, as far as tortious liability is concerned, 'the place of the event giving rise to the damage no less than the place where the damage occurred can, depending on the case, constitute a significant connecting factor from the point of view of jurisdiction'.[18] Thus, the Court added, 'it does not appear appropriate to opt for one of the two connecting factors mentioned to the exclusion of the other, since each of them can ... be particularly helpful from the point of view of the evidence and of the conduct of the proceedings'.[19] On these premises, it concluded that the expression 'place where the harmful event occurred' must be interpreted as meaning that 'the plaintiff has an option to commence proceedings either at the place where the damage occurred or the place of the event giving rise to it'.[20]

In its subsequent case law regarding special jurisdiction over claims in tort, the Court of Justice acknowledged that the principle of ubiquity may give rise to difficulties in particular instances, especially when it comes to defamation and the infringement of personality rights.

---

[18] Case 21-76, *Handelskwekerij G.J. Bier BV v. Mines de potasse d'Alsace SA*, ECLI:EU:C:1976:166, para. 15.
[19] *Ibid.*, para. 17.
[20] *Ibid.*, para. 19.

In *Shevill*, the Court of Justice found that the reasoning underlying *Mines de Potasse d'Alsace* should equally apply, as a matter of principle, in the case of loss or damage other than physical or pecuniary, in particular where reparation is sought for injury to the reputation and good name of a natural or legal person due to a defamatory publication. It held that in the case of a libel by a newspaper article distributed in several states, the place of the event giving rise to the damage 'can only be the place where the publisher of the newspaper in question is established, since that is the place where the harmful event originated and from which the libel was issued and put into circulation': the court of the place where the publisher of the defamatory publication is established must consequently be deemed to have jurisdiction 'to hear the action for damages for *all* the harm caused by the unlawful act'.[21]

As an alternative to bringing proceedings before the latter court (which will normally coincide with the court having general jurisdiction under Article 4(1) Brussels I *bis* Regulation), the plaintiff should be able to sue the defendant in the place where the damage occurred, that is 'the place where the event giving rise to the damage, entailing tortious, delictual or quasi-delictual liability, produced its harmful effects upon the victim'. The Court clarified that in the case of libel in the press, the injury caused by defamation 'occurs in the places where the publication is distributed, when the victim is known in those places', but only to add that the courts of each State in which the publication was distributed may claim jurisdiction only 'to rule on the injury caused *in that State* to the victim's reputation'.[22] This approach, known as the mosaic principle, has been justified by the Court of Justice based on the principle of proximity and on the requirement of the sound administration of justice: as a matter of fact, the courts of each country where the publication has been distributed, the Court observed, 'are territorially the best placed to assess the libel committed in that State and to determine the extent of the corresponding damage'.[23]

More recently, in *eDate*, the Court acknowledged that the principles elaborated in its previous case law regarding the localisation of torts required to be adapted to the peculiar features of injuries consisting in the placing of content online (defamatory content or content infringing an individual's right to respect for his or her private life), as opposed to the physical distribution of media such as printed matter. Online content may in fact be 'consulted instantly by an unlimited number of Internet users throughout the world, irrespective of any intention on the part of the person who placed it in regard to its consultation beyond that person's Member State of establishment and outside of that person's

---

[21] Case C-68/93, *Fiona Shevill, Ixora Trading Inc., Chequepoint SARL and Chequepoint International Ltd v. Presse Alliance SA*, ECLI:EU:C:1995:61, para. 25 (emphasis added).
[22] *Ibid.*, para. 30 (emphasis added).
[23] *Ibid.*, para. 31.

control'.[24] Actually, the Court noted, the Internet reduces the usefulness of the criterion relating to distribution, in so far as the scope of the distribution of content placed online is in principle universal. Furthermore, it may be impossible, on a technical level, to quantify that distribution with certainty and accuracy in relation to a particular State and assess on that basis the damage caused exclusively within the State in question.

Taking these peculiarities into account, the Court held that 'a person who has suffered an infringement of a personality right by means of the Internet may bring an action in one forum in respect of all of the damage caused, depending on the place in which the damage caused in the European Union by that infringement occurred'.[25] Specifically, an action to this effect can be brought before 'the court of the place where the alleged victim has his centre of interests', since this court is the court best placed to assess the impact which online material is liable to have on the personality rights of the claimant.[26] The centre of interests of a person corresponds 'in general' to the habitual residence of that person, but there may be cases, the Court conceded, where the two notions do not coincide, and the centre of a person's interest is in State other than the State where he resides. A similar finding presupposes that 'other factors, such as the pursuit of a professional activity', establish the existence of a particularly close link between the person in question and such other State.[27]

The principles of interpretation developed by the Court of Justice in *Shevill* and *eDate* are also relevant to identifying the court, or courts, that possess special jurisdiction under the Brussels I *bis* Regulation to rule over a claim based on the infringement of the rights of a data subject, namely through the unlawful collection or processing of personal information.[28] The situation of a data subject seeking compensation for the illegal transfer of personal data or the unauthorised disclosure thereof, does not substantially differ from the situation of a victim of online defamation. Proceedings based on similar infringements may accordingly be brought, for the whole damage, before the courts for the place where the data controller is established, or the courts for the place where the data subject has his or her centre of interests.

24　Joined Cases C-509/09 and C-161/10, *eDate Advertising GmbH v. X* and *Olivier Martinez, Robert Martinez v. MGN Limited*, ECLI:EU:C:2011:685, para. 45.
25　*Ibid.*, para. 48.
26　*Ibid.*
27　*Ibid.*, para. 49.
28　See generally B. HESS, 'The Protection of Privacy in the Case Law of the CJEU' in B. HESS and C. MARIOTTINI (eds.), *Protecting Privacy in Private International and Procedural Law and by Data Protection*, Ashgate, Farnham 2015, p. 81. On differences and analogies between defamation and the infringement of privacy rights, see L. GILLIES, 'Jurisdiction for Cross-Border Breach of Personality and Defamation: *eDate Advertising* and *Martinez*' (2012) 61 *International and Comparative Law Quarterly* 1007, 1008.

It is worth observing, on a different (and more general) note, that the rule of special jurisdiction for torts in Article 7(2) Brussels I *bis* Regulation, in addition to being available to the victim of a tort, may also be invoked by the alleged author of the tort, should he wish to obtain a declaration that no tort or delict has been committed, i.e. a declaration of non-liability. The Court found in *Folien Fischer* that the rule now contemplated in Article 7(2) Brussels I *bis* Regulation does not pursue the objective of offering one of the litigants special protection, as is the case of the rules of jurisdiction that apply to proceedings concerning insurance, consumer and employment matters. Noting that the goals pursued by Article 7(2) – proximity and predictability – are not connected either to the allocation of the respective roles of claimant and defendant or to the protection of either, the Court of Justice rejected the view that the application of the rule of special jurisdiction for torts is contingent upon the potential victim initiating proceedings.[29]

It is not clear whether the principle of interpretation in *Folien Fischer* actually applies to any claim within the scope of Article 7(2), including the particular kind of claims considered in *eDate*.[30] The rule developed by the Court in the latter judgment reflects, among other policy considerations, a concern for the effective protection of the victim of online torts. The centre-of-interests criterion reflects, in the Court's reasoning, the 'serious nature of the harm which may be suffered by the holder of a personality right who establishes that information injurious to that right is available on a world-wide basis'.[31] If Article 7(2) Brussels I *bis* Regulation were to be regarded as pursuing a protective goal, if only in respect of violations of personality rights committed through the Internet,[32] the argument put forward by the Court in *Folien Fischer* to reach its conclusion on actions for negative declarations would lose, in this particular area of tort law, some of its weight.

---

[29] Case C-133/11, *Folien Fischer AG and Fofitec AG v. Ritrama SpA*, ECLI:EU:C:2012:664, paras. 45–47. The Court clarified that the issue of jurisdiction is a separate issue from the issue of the admissibility of a negative declaratory action: the latter must be addressed only if jurisdiction is found to exist in the circumstances, and depends (not on regional rules, but) on the applicable national law.

[30] See also, on this topic, M. GEBAUER, 'Persönlichkeitsrechtsverletzung durch Suchererängänzungsfunktion bei Google' (2014) 34 *IPRax* 513, 519 f.

[31] Case C-133/11, *Folien Fischer AG and Fofitec AG v. Ritrama SpA*, ECLI:EU:C:2012:664, para. 47.

[32] Cf, inter alia, H. GAUDEMET-TALLON, *Compétence et exécution des jugements en Europe*, 5th ed., LGDJ, Paris 2015, p. 287 et seq. For a critique of the subjective approach followed by the Court of Justice in *eDate*, see S. SCHMITZ, 'From Where Are They Casting Stones? Determining Jurisdiction in Online Defamation Claims' (2012) 6 *Masaryk University Journal of Law and Technology* 159, 168 et seqq.

3.4. SPECIAL JURISDICTION OVER DISPUTES ARISING OUT OF THE OPERATION OF AN ESTABLISHMENT

Article 7(5) Brussels I *bis* Regulation institutes a special head of jurisdiction for disputes 'arising out of the operations of a branch, agency or other establishment'. Proceedings in connection with these dealings may be brought, if the plaintiff so chooses, before 'the courts for the place where the branch, agency or other establishment is situated'.

For the purposes of Article 7(5), the term 'establishment' refers to a place of business which has the appearance of permanency, has a management, and is materially equipped to negotiate business with third parties.[33] In practice, an establishment may be understood as an entity capable of being the primary, or even exclusive, interlocutor for third parties.[34]

As clarified by the Court of Justice, Article 7(5) may be relied upon to assert jurisdiction over claims of a different nature, provided that the subject matter of the dispute is connected with the operation of the branch or establishment which serves as a head of jurisdiction in the circumstances. These claims, the Court observed, include claims 'concerning non-contractual obligations', as long as they arise from 'the activities in which the branch, agency or other establishment … has engaged at the place in which it is established on behalf of the parent body'.[35] Based on the latter finding, Article 7(5) allows a data subject to sue a controller domiciled in a Member State before the courts for the place where the controller in question has an establishment, provided that the infringement that the claimant complains of originates from the operation of that particular establishment.

3.5. JURISDICTION OVER CO-DEFENDANTS

Article 8(1) Brussels I *bis* Regulation provides that a person domiciled in a Member State, 'where he is one of a number of defendants', may also be sued in the courts of the place where any one of them is domiciled. However, for litigation involving several defendants to be concentrated in the State where one of them is domiciled, the claims brought against each defendant must be 'so closely connected that it is expedient to hear and determine them together to avoid the risk of irreconcilable judgments resulting from separate proceedings', as may occur, *inter alia*, where connected claims are brought against a company

---

[33] Case 33/78, *Somafer SA v. Saar-Ferngas AG*, ECLI:EU:C:1978:205, para. 12.
[34] P. MANKOWSKI, 'Article 7' in U. MAGNUS and P. MANKOWSKI (eds.), *Brussels I bis Regulation*, Köln, Otto Schmidt, 2016, p. 121, 351.
[35] Case 33/78, *Somafer SA v. Saar-Ferngas AG*, ECLI:EU:C:1978:205, para. 13.

and its shareholders, the main debtor and the guarantor, or several participants in a chain of contracts.[36]

In tortious cases, including cases involving the infringement of personality rights, the requirement of connectedness may be met, in principle, where the alleged harm was committed by two or more perpetrators.[37]

3.6. EXPLICIT AND TACIT PROROGATION OF JURISDICTION

Under Article 25 Brussels I *bis* Regulation, if the parties, 'regardless of their domicile, have agreed that a court or the courts of a Member State are to have jurisdiction to settle any disputes which have arisen or which may arise in connection with a particular legal relationship, that court or those courts shall have jurisdiction'. The 'particular legal relationship' referred to in the provision may be a non-contractual relationship arising out of a tort, including, as the case may be, the infringement of personality rights. Logically, a choice-of-court agreement may refer to disputes that have not yet arisen at the time of the agreement itself only if the parties are bound by (or about to enter into) a legal relationship in the framework of which they sense that a particular harmful event is likely to occur. The described scenario, it is submitted, may occur, for example, where an agreement is concluded between a data subject and a data controller in respect of such disputes as may relate to the collection or processing of the personal information provided by the former.

For a choice-of-court agreement to be validly concluded, the parties must comply with the formal requirements set out by the Regulation. The agreement will not be upheld if it is null and void as to its substantive validity under the law of the Member State whose courts have been designated by the agreement itself.

If the agreement is valid, the chosen court is vested, as a rule, with an exclusive jurisdiction to hear the case. As a consequence, all other courts, including the courts designated under Articles 4(1), 7(2) or 8(1) Brussels I *bis* Regulation, must dismiss any claim they might be seised of, if the claim comes with the scope of the agreement. The parties are free to resort to Article 25 to confer non-exclusive jurisdiction to one or more courts of their choice. However, if they intend to do so, they must make it clear in the agreement.

In addition to 'explicit' prorogation of jurisdiction, which has just been described, the Regulation contemplates 'tacit' prorogation. Article 26 provides that jurisdiction lies with the court of the Member State before which the defendant has entered an appearance, apart from where appearance was entered to contest the jurisdiction. Tacit prorogation may occur in respect of proceedings

---

[36] H. Muir Watt, 'Article 8' in U. Magnus and P. Mankowski (eds.), *supra* n. 34, p. 369, 384.
[37] Cf. Case C-228/11, *Melzer v. MF Global UK Ltd*, ECLI:EU:C:2013:305, para. 39.

regarding any claim, with the sole exception of claims for which Article 24 of the Regulation makes provision for 'objective' exclusive jurisdiction.

3.7. JURISDICTION TO ORDER PROVISIONAL MEASURES

The court possessing jurisdiction under the Brussels I *bis* Regulation to decide the merits of a dispute is also entitled, as such, to order such provisional measures as may prove necessary to safeguard the rights of a party during the time needed to adjudicate the matter.

Article 35 Brussels I *bis* Regulation further allows any interested party to apply 'to the courts of a Member State for such provisional, including protective, measures as may be available under the law of that Member State, even if the courts of another Member State have jurisdiction as to the substance of the matter'. In this case, however, a connection must exist between the measure in question and the State before whose courts the application is made. The latter requirement is met, *inter alia*, when the measures are meant to be enforced in the State in question.

The Regulation fails to provide an explicit definition of the expression 'provisional measures'. The formula is understood to refer, generally, to measures aimed at preserving a factual or legal situation as long as proceedings are pending relating to the substance of the dispute. The notion accordingly includes, for example, freezing injunctions, since the aim of these orders is to prevent a party from disposing of or dealing with certain of its assets, as a means to safeguard the requesting party's expectation to actually receive monies it is allegedly entitled to (e.g. as a compensation for the injuries suffered). The jurisdiction of a court to grant provisional measures is similarly at stake where a party seeks an order to prevent the other party from transferring or further processing personal data the collection of which, according to the requesting party, has taken place in violation of the applicable provisions.

On a different note, recital 25 Brussels I *bis* Regulation clarifies that the notion or provisional measures includes 'protective orders aimed at obtaining information or preserving evidence'. However, following the indication of the Court of Justice in *St. Paul Dairy*,[38] measures aimed at obtaining evidence merely as an attempt to decipher whether it is worth bringing an action on the merits should rather fall outside the scope of the rule.[39]

For a measure to be characterised as 'provisional' within the meaning of the Brussels I *bis* Regulation, it is immaterial whether such measure has been applied for in the course of the proceedings or prior to their institution.

---

[38] Case C-104/03, *St. Paul Dairy Industries NV v. Unibel Exser BVBA*, ECLI:EU:C:2005:255.
[39] See further NUYTS, 'Provisional measures' in A. DICKINSON and E. LEIN (eds.), *The Brussels I Regulation Recast*, Oxford University Press, Oxford 2015, p. 357, 368 et seq.

3.8. LIS PENDENS

Alongside jurisdiction, the Brussels I *bis* Regulation addresses the issue of parallel, or duplicate, litigation. 'Parallel' proceedings are proceedings in respect of identical or connected matters that are pending simultaneously before the courts of two or more countries. The Regulation strives to ensure some form of coordination between such proceedings, in particular to avoid the risk of divergent or irreconcilable judgments.

Article 29 is among the provisions of the Regulation entrusted with this task. It is specifically concerned with intra-European *lis pendens*, i.e. the situation that occurs 'where proceedings involving the same cause of action and between the same parties are brought in the courts of different Member States'. In this scenario, according to Article 29(1), 'any court other than the court first seised shall of its own motion stay its proceedings until such time as the jurisdiction of the court first seised is established'. If the court first seised finds that it actually has jurisdiction over the matter, then, pursuant to Article 29(3), any other court must decline jurisdiction in favour of the court first seised.

Article 30 lays down a partly similar provision as regards 'related actions pending in the courts of different Member States'.

For Articles 29 and 30 to apply it is sufficient that the proceedings in question come with the scope of the Brussels I *bis* Regulation. It is immaterial whether the courts seised are to ascertain their jurisdiction in accordance with the uniform rules of jurisdiction set out in the Regulation, or on such other rules on jurisdiction as may be applicable in the circumstances (i.e. the national rules referred to in Article 6, or the EU rules relating to 'specific matters' contemplated in Article 67).

4. JURISDICTION OVER THE INFRINGEMENT OF PRIVACY RIGHTS PURSUANT TO ARTICLE 79(2) GDPR

4.1. THE SCOPE OF APPLICATION OF ARTICLE 79(2) GDPR

The application of Article 79(2) GDPR depends, in principle, on the same conditions that result in the applicability of the GDPR taken as a whole. For these conditions, regard must be had to Articles 1, 2 and 3 GDPR, which define the subject matter of the Regulation and the material and territorial scope of its provisions.

More specifically, Article 79(2) is meant to apply where civil proceedings are brought against a data controller before the courts of a Member State in connection with the infringement of a right protected under the GDPR. For the most part, as suggested by Article 82(6), these will be proceedings instituted by the data subject (or by a body or organisation entitled to do so under Article 80(1) or (2) GDPR) to seek compensation for the damage resulting from the unlawful collection or processing of personal information. However, there seems to be no reason to exclude the applicability of Article 79(2) to proceedings against data controllers other than proceedings aimed at obtaining pecuniary compensation, such as proceedings seeking a mere judicial determination as to the unlawful character of the collection or processing of the claimant's personal data, or proceedings to obtain an injunction to discontinue the processing of certain information or not to disclose or transfer such information to a third party. It is true that the practical result pursued through the latter kind of proceedings may in principle be more easily achieved under a measure issued by the competent supervisory authority in the exercise of the corrective powers now contemplated in Article 58(2)(f) GDPR. However, situations might exist where the data subject has an interest in seeking a judicial remedy, as opposed to (or in combination with) the order of a supervisory authority.[40]

On a different note, the instrumental character of Article 79(2) is relevant to determining the temporal scope of the rule. The GDPR applies, as a whole, 'from 25 May 2018'. Based on the general principle whereby jurisdiction must be assessed in conformity with the rules in force in the forum at the time when the proceedings have been instituted, one might think that Article 79(2) is set to apply to any proceedings initiated on or after the latter date, regardless of when the cause of action arose. However, since the purpose of the provision is to grant an effective judicial remedy to data subjects who consider that their 'rights under [the] Regulation' have been infringed, the applicability of Article 79(2) further presupposes that the infringement itself occurred on or after 25 May 2018.

4.2. THE PROTECTIVE POLICY UNDERLYING ARTICLE 79(2) GDPR

The goal of Article 79(2) GDPR is to strengthen the protection of the rights of data subjects. The rule, it is submitted, rests on the idea that, generally speaking, an imbalance exists between the data subject and the controller. One way to fix, or mitigate, the imbalance is to facilitate the exercise, by data subjects, of their right to a court.

---

[40] For example, the data subject may want to be able to rely on the order issued in a Member State before the authorities of a third country. Depending on the rules in force in the third country in question, this may be easier to achieve if the order is judicial, rather than administrative, in nature.

Three closely inter-related elements indicate that Article 79(2) reflects a concern for data subjects, and actually regards them as weaker parties, in a way that is similar to the approach followed by the Brussels I *bis* Regulation with respect to litigation involving consumers, employees, policy holders and the beneficiaries of an insurance contract.

To begin with, the provision only refers to proceedings against controllers (or processors). The focus is on the interests and expectations of one of the litigants alone: the data subject.

Secondly, Article 79(2) makes provision for two concurrent heads of jurisdiction, thereby granting the claimant the possibility of choosing, in principle, among the courts of two different Member States.[41] Freedom of choice, if properly used, allows the claimant to bring proceedings in the country where his or her rights are likely to enjoy, comparatively, the best protection. It is worth observing in this connection that the factual and legal implications of bringing proceedings in one Member State rather than another may be rather significant in practice. Apart from differences in procedural law (and in terms of efficiency of the judiciary), the substance itself of the rights of the data subject may vary, to some extent, depending on the Member State whose courts are seised of the matter. Though the GDPR introduces, in Article 82, a body of uniform provisions to govern the liability of data controllers and data processors substantively, the regime thus elaborated is far from being complete.[42] The gaps are to be filled by the domestic rules of the law specified under the conflict-of-law provisions of the forum, which vary themselves from one Member State to another.[43]

Finally, among the concurrent heads of jurisdiction provided for under Article 79(2) GDPR is the habitual residence of the data subject. This in fact permits the data subject to seise the courts that should normally be, for him, the easiest to access. The costs associated with the internationality of the dispute would in fact be borne entirely by the defendant.

---

[41] It is worth noting, however, that the options granted to the data subject under Article 79(2) are less numerous than those contemplated in Article 77(1) GDPR for the purpose of determining the Member State before whose supervisory authority the data subject himself is entitled to lodge a complaint. The latter provision actually refers, 'in particular', to the Member State of the habitual residence or place of work of the data subject, as well as to the Member State in which the alleged infringement took place.

[42] Article 82 GDPR is silent, for example, as regards the criteria to be followed for the assessment of damage, and as regards the rules of prescription and limitation.

[43] The conflict-of-law regime of torts has been unified in the EU under Regulation (EC) No. 864/2007 of 11 July 2007 on the law applicable to non-contractual obligations (Rome II) [2007] OJ L199. The Regulation, however, pursuant to Article 1(2)(g), does not apply to 'non-contractual obligations arising out of violations of privacy and rights relating to personality, including defamation'.

4.3. THE JURISDICTION OF THE COURTS OF THE MEMBER STATE WHERE THE ESTABLISHMENT OF THE CONTROLLER IS SITUATED

As mentioned above, Article 79(2) GDPR provides, as a first option, that controllers may be sued before the courts of the Member State where they have 'an establishment'.

It is worth noting at the outset that, unlike Article 7(2) Brussels I *bis* Regulation, which refers to the 'courts for the *place* where the harmful event occurred or may occur', Article 79(2) GDPR confines itself to identifying the Member State whose authorities may be considered, collectively, to possess jurisdiction over a matter. As confirmed by the reference to national law in Article 82(6) GDPR, it is for the law of the designated Member State to set out the criteria for identifying, once jurisdiction has been established, the possible venue of proceedings.

The word 'establishment' must be given an autonomous interpretation. If the expression were to be understood as involving a reference to the notion of establishment employed in the law of a particular Member State, the uniform operation of the Regulation would be jeopardised, and legal certainty would be at risk.

Article 4 GDPR, which provides a list of definitions for the purposes of the GDPR, clarifies how the expression 'main establishment' should be understood, but does not include, strictly speaking, a definition of 'establishment'.[44] For the latter expression, regard should be had to recital 22. This states that an establishment 'implies the effective and real exercise of activity through stable arrangements' and that '[t]he legal form of such arrangements, whether through a branch or a subsidiary with a legal personality, is not the determining factor in that respect'.[45]

It is clear from the foregoing that, under Article 79(2), proceedings against a controller do not have to be brought before the courts of the Member State where

---

[44] According to Article 4(16)(a), the main establishment of a controller, namely one 'with establishments in more than one Member State', is 'the place of its central administration in the Union, unless the decisions on the purposes and means of the processing of personal data are taken in another establishment of the controller in the Union and the latter establishment has the power to have such decisions implemented, in which case the establishment having taken such decisions is to be considered to be the main establishment'. For its part, Article 4(16)(b) provides that the main establishment of a processor having establishments in more than one Member State is 'the place of its central administration in the Union, or, if the processor has no central administration in the Union, the establishment of the processor in the Union where the main processing activities in the context of the activities of an establishment of the processor take place to the extent that the processor is subject to specific obligations' under the GDPR.

[45] On the notion of 'establishment', see also *Opinion 8/2010 on Applicable Law*, adopted by the Article 29 Data Protection Working Party on 16 December 2010, 0836–02/10/EN, WP 179, p. 11 et seq.

the 'main establishment' of the defendant is situated, but may rather be instituted in the Member State where the defendant has a place of business through which it carries out its activity on a permanent basis.

That said, one may wonder whether Article 79(2) GDPR involves the possibility of seising the courts of the Member State in which *any* of the establishments of the controller is situated, or whether the provision should rather be understood as referring to the particular establishment the operation of which has resulted in the infringement that the claimant complains of in the circumstances.

A narrow reading might in principle be argued for based on an analogy with Article 7(2) and (5) Brussels I *bis* Regulation. As clarified by the Court of Justice in *Shevill* as regards the rule now laid down in Article 7(2), the place where the person claimed to be liable is 'established' is, in the case of a libel, the place where the event giving rise to the damage occurred, 'since that is the place where the harmful event originated and from which the libel was issued and put into circulation'. For its part, Article 7(5), as already observed, is explicit in specifying that the jurisdiction of the courts for the place of a branch or establishment is limited to such claims as appear to be connected with the particular establishment in question.

It is contended that, in spite of the foregoing, a broader reading should be preferred. Under this broader reading, the claimant should be allowed to rely on Article 79(2) GDPR without having to show a nexus between the alleged infringement and the operation of the establishment to which reference is made for the purposes of jurisdiction, and that he should be allowed to do so to have a determination as to all the consequences of the infringement, no matter where they might have been suffered.

Two arguments, one literal, the other teleological, support this view. On the one hand, Article 79(2), by making use of an indefinite article ('*an* establishment', '*un* établissement', '*eine* Niederlassung', '*uno* stabilimento'), suggest that the provision does not actually intend to single out a particular establishment of the controller for the purpose of identifying the courts with jurisdiction. On the other hand, if the data subject were to provide evidence of a connection between the cause of action and the establishment in question as a prerequisite for claiming the jurisdiction of the courts of the State of the establishment, Article 79(2) would lose much of its practical usefulness, and ultimately fail to achieve its goal of providing the data subject with a special protection. Actually, due to the peculiar nature of data (which are immaterial and ubiquitous), it could prove extremely difficult for the claimant to demonstrate that the infringement at issue originated from a particular place, instead of another. It should also be borne in mind that, as the Court of Justice noted in *Google Spain* with reference to Article 4(1)(a) of Directive 95/46,[46] the activities of a controller may be inextricably linked to those of its establishments. In these circumstances, a clear distinction between the dealings of the main establishment and those of the other establishments, no less than a clear distinction among the dealings of the various establishments of a single controller, may prove impossible to make.

It is true that the suggested broad reading may prompt the claimant to cherry-pick from among the various establishments of the defendant the one situated in the Member State whose courts would provide him the best chances with respect to litigation. It should be noted, however, that this strategy appears to be in line with the protective policy underlying Article 79(2) GDPR, and that the extensive substantive harmonisation achieved by the GDPR significantly reduces the impact of similar litigation tactics. At any rate, it is submitted that the claimant should be barred from resorting to cherry-picking only if it is proved, to the extent required under the general prohibition of abuse of rights in EU law, that he is aware that the selected establishment has in no way been involved in the collection or processing of his personal information, and that his choice is but a means to achieve an advantage that he would normally not be entitled to.[47]

4.4. JURISDICTION AT THE HABITUAL RESIDENCE OF THE DATA SUBJECT

The second option envisaged in Article 79(2) GDPR consists in the possibility for the data subject to sue the data controller before the courts of the Member State of his or her habitual residence. Here, too, the provision confines itself to determining the State whose courts, globally considered, possess jurisdiction, and is not concerned with the issue of venue.

Habitual residence is not defined in the GDPR. Several EU legislative measures make use of this expression (in particular as a connecting factor for the purpose of deciding conflicts of laws and conflicts of jurisdiction), but similarly fail to include an explicit definition of the concept, or merely provide hints as to way in which habitual residence should be assessed in particular circumstances for the purposes of the provisions therein.[48]

---

[46] Case C-131/12, *Google Spain SL, Google Inc. v. Agencia Española de Protección de Datos (AEPD), Mario Costeja González*, ECLI:EU:C:2014:317, para. 56.

[47] On the scope and operation of the prohibition of abuse of rights as regards the rules governing the jurisdiction of courts, see generally L. USUNIER, 'Le règlement Bruxelles I *bis* et la théorie de l'abus de droit' in E. GUINCHARD (ed.), *Le nouveau règlement Bruxelles I bis*, Bruylant, Brussels 2014, p. 449.

[48] See, for instance, recitals 23 and 24 of Regulation (EU) No. 650/2012 of 4 July 2012 on jurisdiction, applicable law, recognition and enforcement of decisions and acceptance and

Generally speaking, habitual residence is understood by the Court of Justice to refer to place where a person has 'established, with the intention that it should be of a lasting character, the permanent or habitual centre of his interests'.[49] A variety of elements may need to be considered to identify the habitual residence of a person, such as the family situation of the individual in question, the length and continuity of his or her residence in a given place, the fact that he or she is in stable employment and, if he or she has moved from one place to another, the reasons which have led him or her to move.[50] In practice, both objective circumstances and the intentions of the person concerned have to be examined and assessed. The statements made by the interested person as to his or her habitual residence do carry some weight, but they are not sufficient, as such, to decide the issue of habitual residence. The notion is actually meant to reflect the actual integration of the individual in a given social environment, and cannot accordingly depend on a mere declaration of the person in question or on formal or official qualifications.

By its nature, the habitual residence of a person may change over time. For a change of habitual residence to become effective, no requirement of a minimum stay in the new residence is established. To determine whether the habitual residence of the individual in question has changed, one merely has to (re)assess the elements indicated above to see whether the centre of the interests of the individual has moved from one place to another. The view is generally accepted that a person can have but one habitual residence at a time.

The reference made by Article 79(2) GDPR to the habitual residence of the data subject echoes the centre-of-interests rule developed by the Court of Justice in *eDate*. Here, too, the idea is that the victim should be allowed to have a determination made by one court of the damage suffered globally as a result of the infringement of his or her personality rights, and that the court in question should be identified by looking at the country where the core of the personal, economic and social interests of the victim appears to be situated.

While the analogy with the *eDate* scenario appears to be justified in general terms, the transposition of the centre-of-interests rule from libel to data protection is likely to prove problematic in at least one respect. In *eDate*, the Court observed that the jurisdiction of the court of the place where the victim has the centre of his or her interests 'is in accordance with the aim of predictability of the rules governing jurisdiction … also with regard to the defendant, given that the publisher of harmful content is, at the time at which that content is placed online, in a position to know the centres of interests of the

---

enforcement of authentic instruments in matters of succession and on the creation of a European Certificate of Succession [2012] OJ L201.

[49] See in these terms, for example, Case T-298/02, *Herrero Romeu v. Commission*, ECLI:EU:T:2005:369, para. 51.

[50] Case C-90/97, *Robin Swaddling v. Adjudication Officer*, ECLI:EU:C:1999:96, para. 29.

persons who are the subject of that content'.⁵¹ Now, unlike the publication of a defamatory article, which may in fact (and should) be preceded by some form of investigation into the person to which the article relates, including that person's whereabouts and interests, the infringement of the rules on the protection of personal data may plausibly occur, in some circumstances, without the controller being given a similar prior opportunity to determine the habitual residence of the victim (as could be the case, for example, of the accidental destruction or unauthorised disclosure of personal data, before the processing has been carried out). Admittedly, the adoption of properly conceived collection and processing protocols, involving as the case may be the use geolocation technologies, should provide a partial answer to these concerns. Yet one may wonder whether the jurisdiction conferred under Article 79(2) GDPR to the courts of the Member State where the data subject resides would in every instance be consistent with the principle of predictability, the respect for which is crucial – as observed above – to ensuring, at the level of the rules on jurisdiction, the fairness of civil proceedings.

5. THE COORDINATION BETWEEN ARTICLE 79(2) GDPR AND THE BRUSSELS I *BIS* REGULATION

5.1. ASSESSING WHETHER THE APPLICATION OF A PROVISION IN THE BRUSSELS I *BIS* REGULATION WOULD BE PREJUDICIAL TO THE GDPR

Article 79(2) GDPR does not rule out entirely the concurrent application of the general regime of the Brussels I *bis* Regulation. The problem accordingly arises of precisely identifying the cases in which a court sitting in a Member State should be allowed to resort to the latter Regulation as a means to supplement the special rule of jurisdiction in the GDPR.

Both the Brussels I *bis* Regulation and the GDPR content themselves with saying – respectively in Article 67 and in recital 147, as already noted – that the former does not 'prejudice' the latter.

The expression, it is submitted, reflects a concern for the effectiveness of Article 79(2) GDPR. A prejudice should thus be deemed to exist within the meaning of the mentioned provisions when, and to the extent to which, the application of a particular provision of the Brussels I *bis* Regulation affects the ability of the rule in the GDPR to achieve its purposes.

Specifically, a provision in the Brussels I *bis* Regulation should be regarded as prejudicial to the GDPR not only if it prevents the realisation of a practical

---

51   Joined Cases C-509/09 and C-161/10, *eDate Advertising GmbH v. X* and *Olivier Martinez, Robert Martinez v. MGN Limited*, ECLI:EU:C:2011:685, para. 50.

outcome contemplated under Article 79(2) GDPR, but also if it deals with the particular issues that the latter is specifically designed to address under policy considerations that do not coincide with those underlying the GDPR. Put otherwise, the Brussels I *bis* Regulation must be disregarded where its application would either contradict Article 79(2) GDPR or affect the integrity and consistency of the special regime provided therein. If none of these situations occur, nothing should prevent the parties of proceedings of the kind considered in Article 79(2) from relying on the pertinent provisions of the Brussels I *bis* Regulation.

5.2. GENERAL RULES THAT DO NOT APPEAR TO PREJUDICE ARTICLE 79(2) GDPR

Based on the foregoing, it is believed that the operation of Article 79(2) GDPR would in no way be prejudiced by the application of Article 4(1) Brussels I *bis* Regulation, i.e. the rule that confers general jurisdiction over civil and commercial matters to the courts of the Member State of the domicile of the defendant. Admittedly, the finding is of little practical importance as regards proceedings against a data controller, since the domicile of the latter, for the purposes of Article 4(1), will almost invariably coincide with an 'establishment' of the latter for the purposes of Article 79(2) GDPR.[52]

No prejudice, it is submitted, would be suffered by the GDPR from the application of Article 8(1) Brussels I *bis* Regulation, concerning jurisdiction over co-defendants. The data subject should thus be allowed to sue before the same court two or more controllers who are neither domiciled nor established in the same Member State, if he or she considers, for example, that such defendants are jointly liable for the particular infringement he complains of. Indeed, a concentration of claims would already be possible, pursuant to Article 79(2) GDPR, before the courts of the Member State where the data subject habitually resides. Article 8(1) Brussels I *bis* Regulation provides the data subject with one more opportunity to concentrate the dispute in one court, namely the court for the place where any one of the co-defendants is domiciled. The outcome is in keeping with the protective purpose of Article 79(2) GDPR and does not appear to be unfair to the defendants, since, in this scenario, jurisdiction over the various defendants would ultimately rest on a rule of general application.

The rules of the Brussels I *bis* Regulation that concern jurisdiction to order provisional measures (as may be requested, for example, by the data subject with

---

[52] For similar, if not stronger, reasons it is of little practical importance to determine whether Article 7(5) Brussels I *bis* Regulation can still be relied upon for the purpose of proceedings within the scope of Article 79(2) GDPR, since both provisions refer to an 'establishment' of the defendant for the purpose of identifying the court with jurisdiction (albeit, as seen, with slightly different implications).

a view to obtaining evidence to substantiate a claim against a controller) are equally unable, it is submitted, to prejudice the operation of Article 79(2) GDPR. These rules, too, should accordingly be deemed to be applicable to the set of proceedings considered in the latter provision.

5.2. GENERAL RULES THAT APPEAR TO PREJUDICE ARTICLE 79(2) GDPR

The criteria set out above to assess the prejudicial or non-prejudicial character of the provisions in the Brussels I *bis* Regulation, suggest, by contrast, that there should be no room left for Article 7(2) of that Regulation as regards the kind of proceedings referred to in Article 79(2) GDPR.

The latter provision, as observed earlier, builds to a large extent on the principles of interpretation elaborated by the Court of Justice in respect of the rule of special jurisdiction in matters relating to torts or delict that is presently featured in Article 7(2) Brussels I *bis* Regulation. Now that these principles have been refashioned through the GDPR and adapted to the peculiar features of the law of data protection, the application of the rule that served as a model for these developments (as interpreted by the Court of Justice, *inter alia*, in *Shevill*, *eDate* and *Folien Fischer*) would harm the clarity and unity of the special regime laid down in Article 79(2) GDPR, and would thus prove prejudicial to the latter.[53]

5.3. GENERAL RULES WHOSE APPLICABILITY TO PROCEEDINGS ENVISAGED IN ARTICLE 79(2) GDPR APPEARS TO BE UNCERTAIN

Doubts may be expressed as to whether, and in what circumstances, the ongoing application of the rules of the Brussels I *bis* Regulation regarding parallel proceedings and prorogation of jurisdiction could actually be regarded as prejudicial to Article 79(2) GDPR.

---

[53] It is worth observing that, while jurisdiction under Article 7(2) Brussels I *bis* Regulation depends on the localisation of the *subject matter* of the dispute (through the localisation of a particular aspect thereof: the harmful event), Article 79(2) GDPR makes jurisdiction contingent on the localisation of the *parties* to the dispute. It is true that the reference to (the centre of interests and) the habitual residence of the data subject builds on the idea that the harm suffered by the victim tend to concentrate *objectively* in that place, but the shift from a purely objective ground of jurisdiction to a ground of jurisdiction that presents itself as personal in nature, suggests that the two provisions underlie a largely different logic. It should also be noted, on a more pragmatic note, that the use of a personal connecting factor to localise disputes in matters relating to data protection allows the seised court (and the litigants) to avoid the inherent difficulties surrounding the localisation of online dealings (regarding which see P. MANKOWSKI, 'Article 7', *supra* n. 34, p. 283). If Article 7(2) of the Brussels I *bis* Regulation were to apply to the kind of proceedings considered in Article 79(2) of the GDPR, these practical difficulties would likely reappear.

The problem with *lis pendens* and related actions arises from the fact the GDPR itself includes a provision on parallel proceedings, which *prima facie* leaves no room for such general rules on the same subject matter as may be found in Articles 29 and 30, among others, of the Brussels I *bis* Regulation. Article 81 GDPR, labelled 'Suspension of proceedings', refers among other things to the situation where 'proceedings concerning the same subject matter as regards processing of the same controller or processor' are pending in the courts of two Member States. In this case, according to Article 81(2), 'any competent court other than the court first seized may suspend its proceedings'. Pursuant to Article 81(3), where the proceedings in question 'are pending at first instance, any court other than the court first seized may also, on the application of one of the parties, decline jurisdiction if the court first seized has jurisdiction over the actions in question and its law permits the consolidation thereof'.

The wording of Article 81 seems to suggest that the provision is intended to apply, as such, to all the proceedings that come with the scope of the GDPR. For its part, recital 144 (the only statement in the Preamble devoted to the rule on suspension of proceedings) seems to indicate that Article 81 must be given a narrower reading, as it is merely concerned with 'proceedings against a decision by a supervisory authority'. This indication appears to imply that Article 81 has simply no role to play when it comes to claims of the kind referred to in Article 79(2). If the narrower reading is correct, then nothing should prevent the rules of the Brussels I *bis* Regulation on parallel proceedings from being applied to the latter kind of proceedings. Though the wording of Article 81 is, at best, ambiguous, the narrower reading would appear to be the most preferable interpretive option. As a matter of fact, a different interpretation would be difficult to justify. The approach of the GDPR to parallel proceedings is definitely less sophisticated than the approach of the general regime on *lis pendens* as set out in the Brussels I *bis* Regulation. It is actually unclear why the general regime – which is largely known by practitioners and has proved to work well in practice – should be abandoned, in this area, to be replaced by the former.[54]

Doubts of a different nature surround the applicability of the rules laid down in the Brussels I *bis* Regulation to govern choice-of-court agreements and tacit prorogation of jurisdiction, i.e. Articles 25 and 26.

The GDPR does not specify whether jurisdiction under Article 79(2) is exclusive in nature. Nor does it clarify whether, and in what way, the grounds therein may be derogated from under an agreement between the parties.

---

[54] See in this connection P. DE MIGUEL ASENSIO, 'Aspectos internacionales del Reglamento general de protección de datos de la UE (I): cuestiones de competencia' <http://pedrodemiguelasensio.blogspot.it>.

As shown by the rules in the Brussels I *bis* Regulation concerning insurance, consumer and labour disputes, prorogation of jurisdiction is not necessarily incompatible with the protection of a weaker party.[55] True enough, there a cases where the weaker party runs the risk of being deprived, through a choice-of-court agreement, of the jurisdictional advantages that he or she would otherwise enjoy if jurisdiction was established on purely objective grounds. Yet the risk may be minimised, or avoided, by selectively limiting the admissibility of explicit prorogation of jurisdiction, and by creating special duties of information the compliance with which is a prerequisite for the operation of tacit prorogation.[56]

In particular, the rules of the Brussels I *bis* Regulation confine explicit prorogation in these sensitive areas to cases where the agreement itself would be beneficial to the weaker party (namely, where the agreement does not purport to exclude the jurisdiction of the courts designated objectively, but rather extends the lists of the fora available to the weaker litigant), and the situations where the protected party is, in principle, in a position to make a genuine and informed choice on the issue (namely, where the agreement is entered into only after the dispute has arisen).[57]

Based on the protective purpose of Article 79(2) GDPR, one would be tempted to resort to analogy and actually apply to the disputes contemplated in the latter provision (with the necessary adaptations) the rules under which the Brussels I *bis* Regulation allows the parties to derogate by agreement from the objective protective fora instituted for insurance, consumer and employment matters. A similar approach, however, would not be entirely convincing, since the regimes on insurance, consumer and employment disputes are meant to apply to specific classes of claims, and can hardly be regarded as embodying, in the aspects considered, a set of solutions warranting general applicability.

Taking all of the foregoing into account, it is submitted that the general rules of the Brussels I *bis* Regulation on explicit prorogation of jurisdiction should be deemed to prejudice the operation of Article 79(2) GDPR whenever their application would restrict the right of the data subject to sue a controller before

---

[55] The assertion finds further support in the rules on jurisdiction applicable to cross-border disputes over matters of family maintenance, as set out in Regulation (EC) No. 4/2009 of 18 December 2008 on jurisdiction, applicable law, recognition and enforcement of decisions and cooperation in matters relating to maintenance obligations [2009] OJ L7. The Regulation, which is meant to enhance the protection of the maintenance creditor, allows the parties, under Article 4, to agree on the court, or courts, that should have jurisdiction to settle their disputes. The jurisdiction thus conferred on the designated court or courts 'shall be exclusive unless the parties have agreed otherwise'.

[56] See Article 26(2) of the Brussels I *bis* Regulation.

[57] The relevant provisions are to be found in Articles 15, 19 and 23 of the Brussels I *bis* Regulation, respectively.

the courts specified under the latter provision. Conversely, nothing seems to prevent a general rule such as Article 25 of the Brussels I *bis* Regulation from being relied upon to uphold an agreement that, while preserving the operation of Article 79(2) GDPR, would allow the data subject to *also* bring proceedings in the courts of one more Member States other than those provided for under the GDPR.

PART III
THE LEGISLATIVE INSTRUMENTS FOR A DIGITAL SINGLE MARKET

A EUROPEAN MARKET FOR DIGITAL GOODS

Michael Lehmann

1. Digital Goods 112
2. Portability 113
3. The Supply of Digital Content 115
   3.1. Harmonisation 115
   3.2. Definitions 116
      3.2.1. Digital Content 116
      3.2.2. Supply 116
   3.3. Scope of Application 116
      3.3.1. Personal 116
      3.3.2. Substantive 117
      3.3.3. Exceptions 118
   3.4. Relationship between the Directive and Other Regulations 118
   3.5. Supplier's Obligations 119
      3.5.1. Supply of Digital Content 119
      3.5.2. Freedom from Defects 119
      3.5.3. Liability 120
   3.6. Consumer Rights 121
      3.6.1. Non-performance 121
      3.6.2. Defects 121
         3.6.2.1. Termination of the Contract in the Event of Defect 122
         3.6.2.2. Rescission of the Contract 122
         3.6.2.3. Price Reduction 123
         3.6.2.4. Damages 123
      3.6.3. Other Provisions 123
         3.6.3.1. Right to Terminate Long-Term Contracts 123
         3.6.3.2. Right of Redress 124
         3.6.3.3. Enforcement 124
         3.6.3.4. Mandatory Nature 124
   3.7. Conclusion 124
4. Online Trade in Goods 125
5. Summary 126

This article considers the Commission's proposal of three sets of regulations as it moves towards creating a digital internal market. It discusses the draft Regulation on cross-border data portability (section 2), and the draft Directives concerning digital content (section 3) and the online trade in goods (section 4), followed by an initial analysis. The article ends with the opinion (section 5) that the Commission has presented a highly welcome proposal for the creation of a digital internal market that, above all with respect to the field of contracts concerning the provision of digital content, is worthy of support from the point of view of commercial law.

1. DIGITAL GOODS

In his 'government statement' Jean-Claude Junker[1] and, following him, Commissioners G. Oettinger[2] and A. Ansip[3] attached top priority to the commercial law structure of the European digital internal market. Accordingly, the Commission has now presented its ideas for the modernisation of copyright within the EU[4] and as a first step has published three proposals for one Regulation and two new Directives aimed at creating a working digital internal market: a Regulation on ensuring the cross-border portability of online content services in the internal market,[5] a Directive intended to regulate contracts for the supply of digital content,[6] and a proposal for a Directive concerning contracts for the online and other distance sales of goods.[7]

Article 2(11) of the Consumer Rights Directive,[8] which concerns distance sales and the conclusion of off-premises contracts, has already defined digital content as follows: 'Data which are produced and supplied in digital form'. This is explained in further detail in recital 19: 'Data which are produced and supplied in digital form, such as computer programs, applications, games, music, videos or texts, irrespective of whether they are accessed through downloading or streaming, from a tangible medium or through any other means.'

---

[1] COM(2015) 192 final.
[2] Strategie für einen digitalen europäischen Binnenmarkt (2015) 9 *K&R*, Editorial.
[3] (2015) 6 *MMR*, Editorial.
[4] Press release of 09.12.2015 (2016) *EuZW* 3; cf. J. Brauneck (2015) *GRUR Int.* 889 et seq.; J. Reinbothe (2015) *ZGE* 145 et seq.; G. Spindler (2016) *CR* 73 et seq.; see also M. Wendland (2016) *EuZW* 126 et seq.
[5] COM(2015) 627 final.
[6] COM(2015) 634 final.
[7] COM(2015) 635 final.
[8] Directive 2011/83/EU [2011] OJ L304/64; cf. A. De Franceschi and M. Lehmann (2015) *The Italian Law Journal* 51 et seq., 55 et seq.; M. Lehmann, in J. Conrad and M. Grützmacher (eds.), *Recht der Daten und Datenbanken im Unternehmen*, Otto Schmidt, Cologne 2014, p. 134 et seq.

Article 2(j) of the draft Regulation on a Common European Sales Law[9] contains the following definition: 'Data which are produced and supplied in digital form, whether or not according to the buyer's specifications, including video, audio, picture or written digital content, digital games, software and digital content which makes it possible to personalise existing hardware or software ...'.

The intention is to establish a functioning and efficient internal market for these digital goods within the EU with – at least in the long term – no distinction being made as to whether they are supplied, marketed or 'provided' on a physical medium, e.g. a CD-ROM or DVD, or intangibly through the Internet, for instance in the cloud[10] through streaming, in other words by being made available on the Internet and downloaded.

This concept of 'digital content' ought really, and more in conformity with the market, be referred to as 'digital goods' and interpreted accordingly, since what is traded and used commercially is network-compatible 'data records'.[11]

According to the intentions of the Commission, a functioning European internal market is to be created for all digital goods[12] since the fundamental freedoms of the free movement of services and goods (Arts. 56 and 34 TFEU) can be implemented particularly efficiently via the Internet.

Following the failure of the proposal for a Common European Sales Law,[13] new sectoral sub-regulations are to be created that above all could be of value for the trade in digital goods within the EU.

2. PORTABILITY

The draft Regulation[14] on ensuring the cross-border portability of online content services in the internal market involves a reduction of the territorial principle[15] by the home country principle, as already known from the Satellite

---

[9] COM(2011) 635 final, cf. in detail J. DRUSCHEL, *Die Behandlung digitaler Inhalte im Gemeinsamen Europäischen Kaufrecht (GEKR)*, 2014, p. 21 et seq.; J. DRUSCHEL and OEHMICHEN (2015) *CR* 173 et seq., 175 et seq.; *id.* (2015) *CR* 233 et seq.

[10] On cloud computing see G. BORGES and J. MEENTS (eds.), *Cloud Computing*, Beck, Munich 2016, *passim*; A. GIEDKE, *Cloud Computing: Eine wirtschaftsrechtliche Analyse mit besonderer Berücksichtigung des Urheberrechts*, Utz, Munich 2013, *passim*, each with extensive references.

[11] A. OHLY (2013) *JZ* 43; R. HILTY (2012) *CR* 628 et seq.; M. LEHMANN (2015) *GRUR Int.* 678; LEHMANN, *supra* n. 8), p. 134 et seq.

[12] Cf. European Commission press release of 09.12.2015; see also *supra* n. 1: 'A Digital Single Market Strategy for Europe'.

[13] See *supra* n. 9; but cf. also P. SVOBODA, 'The Common European Sales Law – will the Phoenix rise from the Ashes again?' (2015) *ZEuP* 689 et seq.

[14] COM(2015) 627 final (2016) *EuZW* 3 et seq.; cf. also the extensive working document, SWD(2015) 270 final.

[15] Cf. on the problem in the net, LEHMANN in BORGES and MEENTS (eds.), *supra* n. 10, p. 445 et seq.; cf. also N. RAUER and D. ETTIG (2016) *K&R* 79, 81.

Broadcasting and Cable Retransmission Directive.[16] In the same way as telephony roaming within the EU, streaming should also be made possible without borders within this economic area, with the result that for the users there should no longer be any national borders within European cloud services, thus making geo-blocking impossible in the future. Pursuant to Article 3 of the Regulation, anyone who makes available within an EU Member State in which they are domiciled an online content service such as e-books[17] or e-music to subscribers within the framework of service contracts, for instance by means of streaming or other downloading, must enable its customers who are 'temporarily' present in another Member State to access and use the online content services in this state, in order for instance to be able to read an e-book on holiday in another European country.[18] Pursuant to Article 3(2) of the Regulation, the same technical quality as in the country in which they are domiciled need not be ensured. This use, while abroad in Europe, pursuant to Article 4 of the Regulation, deemed solely to occur in the country of residence, and legal regulations, such as national copyright regulations, that are intended to prevent this are, pursuant to Article 5, 'unenforceable'; this also applies to the protection of databases,[19] copyright in the information society,[20] the rental right,[21] and the protection of computer programs[22] and of audio-visual media services.[23] The ECJ had already held similarly in *Football Association Premier League v. Murphy*[24] that national restrictions on the provision of services cannot be justified by the protection of intellectual property and constitute a restraint on competition prohibited by Article 101 TFEU. Pursuant to Article 5(2) of the Regulation, authors and the holders of related rights can verify whether this use is made within the framework of a subscription contract concluded for the country of residence and whether a corresponding licence has also been issued for this use.

---

[16] Directive 93/83/EEC [1993] OJ L248/15, cf. Art. 1(2)(b) thereof: 'The act of communication to the public by satellite occurs solely in the Member State where, under the control and responsibility of the broadcasting organization, the programme-carrying signals are introduced into an uninterrupted chain of communication leading to the satellite and down towards the earth.' See also Art. 3 of the Electronic Commerce Directive 2000/31/EC [2000] OJ L178/1, whose schedule, however, excludes copyright and industrial property rights from the country-of-origin principle. Cf. M. LEHMANN in M. LEHMANN (ed.), *Electronic Business in Europa*, 2002, p. 100 et seq.; F. MAENNEL in M. LEHMANN (ed.), *Electronic Business in Europa*, 2002, p. 52 et seq.; for a differentiated approach see REINBOTHE, *supra* n. 4, p. 164 et seq.

[17] Thus also G. SPINDLER (2016) *CR* 73, 75.

[18] Cf. also *ibid.*, p. 75, who, however, considers that the scope of application goes beyond holiday travel.

[19] Directive 96/6/EC [1996] OJ L77/20.

[20] Directive 2001/29/EC [2001] OJ L167/10.

[21] Directive 2006/115/EC [2006]OJ L376/28.

[22] Directive 2009/24/EC [2009] OJ L111/16.

[23] Directive 2010/13/EU [2010] OJ L95/1.

[24] ECJ, *FAPL/Murphy* (2012) *GRUR* 156 et seq.

As regards services within the framework of a digital subscription service within the EU, the intention is indeed to open up a border-free digital European economic area without any national restrictions, an objective that without doubt is to be assessed as very consumer friendly. The only criticism here is the restriction of portability to a 'temporary presence' in a Member State other than the country of residence of the licence contract. However, this is a first step in the right direction and should in terms of its evolutionary thrust be assessed positively as the start of the development of a functioning European digital internal market.[25]

3. THE SUPPLY OF DIGITAL CONTENT

3.1. HARMONISATION

In contrast to the failed draft Regulation on the introduction of a Common European Sales Law (CESL), the Commission has decided against an optional instrument and in favour of an (in principle) full harmonisation Directive contracts for the supply of digital content[26] (Art. 4 draft Directive).[27] In return, the direct intervention in national contract law is relatively small. There are two reasons for this: firstly, there are hardly any specific national contract law regulations in this sector;[28] and secondly the Directive restricts itself to the B2C sector and to the regulation of the most important contractual entrepreneurial duties, above all concerning freedom from defect, and the consumer's rights in the event of breaches of duty, in particular legal remedies for defects (Art. 1 draft Directive).[29]

Thus the Directive is intended to close gaps concerning contracts for digital content left open above all by the Consumer Goods Directive[30] and the Consumer Rights Directive. While the former fundamentally modernises national contract law, its regulatory scope only extends to purely physical goods and hence at best to embodied digital content. In contrast, the latter was the first to include digital content irrespective of the manner of its delivery,[31] but the Directive's content was restricted to the introduction of information obligations for contracts concerning the supply of digital content and provisions concerning the cancellation right and the consequences of exercising it.

---

25 Similarly RAUER and ETTIG, *supra* n. 15, p. 83, who see the proposal at least as 'a small step towards a cross-border use of European works.'
26 COM(2015) 634 final.
27 Cf. also recital 5 et seq.
28 J. DRUSCHEL and M. OEHMICHEN (2015) *CR* 173, 176; ID. (2015) *CR* 233, 239.
29 Cf. also recital 3.
30 Directive 1999/44/EC [1999] OJ L171/12.
31 Cf. DRUSCHEL, *supra* n. 9, p. 40 et seq.

3.2. DEFINITIONS

The draft Directive is preceded by a total of 11 definitions in the traditional manner of European secondary law.

3.2.1. *Digital Content*

The wording of the definition of digital content is deliberately broad.[32] Alongside the definition covering software and typical digital consumer goods as found in Article 2(11) of the Consumer Rights Directive (cf. Art. 2(1)(a) draft Directive, which speaks of 'data which is produced and supplied in digital form, for example video, audio, applications, digital games and any other software'), Article 1(1)(b) and (c) of the draft Directive goes considerably further by including services allowing the creation, processing or storage of data in digital form, where such data is provided by the consumer (Art. 1(1)(b)), and services allowing sharing of and any other interaction with data in digital form provided by other users of the service (Art. 1(1)(c)). Thus the definition of digital content also covers all forms of cloud computing[33] and social networks.[34]

3.2.2. *Supply*

The English version[35] of the present draft Directive uses the term 'supply'. The German version uses the term *'Bereitstellung'* (furnishing). This latter term appears to do justice to the breadth of the definition. According to Article 2(10) of the draft Directive, it is intended to cover the provision of access to digital content or making digital content available. This broad definition is the result of the Directive's intention to cover all forms of the digital economy.

3.3. SCOPE OF APPLICATION

3.3.1. *Personal*

In personal terms, the draft Directive is restricted to contracts between enterprises and consumers (Art. 1), the former being the party that supplies the digital content and the latter the party that receives it (Art. 3(1)). Thus the provisions apply neither to contracts exclusively between enterprises nor to those between consumers.

---

[32] Recital 11; WENDLAND, *supra* n. 4, p. 126 et seq.
[33] Cf. DRUSCHEL, *supra* n. 9, p. 100 et seq.
[34] Cf. also recital 11.
[35] Cf. *supra* n. 6.

3.3.2. Substantive

In substantive terms, the Directive applies to contracts concerning the supply of digital content. In accordance with the broad definitions of both digital content and its supply, it covers a large number of contractual structures that are encountered on a daily basis. In the light of the variety of the existing business models and the rapid developments in this field, the Commission has deliberately not restricted itself to a classical contract typology (purchase, rent/hire, contract for work and services, licence rights),[36] but instead, independently of the classification of such contracts in national law, regulates entrepreneurial duties and consumer rights in order to eliminate the existing obstacles on the internal market and to create legal certainty.[37] Accordingly, the applicability of the Directive will not depend on whether the consumer pays a sum of money or whether he discloses personal details as consideration: the sole decisive factor is that the transaction is not a gift[38] but instead involves some kind of consideration that the consumer provides for the supply of digital content (Art. 3(1))

In line with the broad definition of digital content and its supply, the Directive is also intended to apply to contracts that provide for the supply of digital content produced according to customer specifications. This includes constellations that in Germany hitherto fell under the contract for work and services (§631 et seq. Civil Code) or contracts for work and materials (§651 Civil Code).[39] As already laid down in the Consumer Rights Directive, and in conformity with the ECJ decision in the *Oracle v. UsedSoft* case,[40] the distinction between embodied and unembodied digital content should as a matter of principle be irrelevant provided that the data carrier only serves as a means of transport. Consequently, Article 20(1) of the Directive amends the Consumer Sales Directive such that its definition of goods also includes non-embodied digital content if a hardware medium is only used as carrier.[41]

This approach is convincing, since the subject matter of the contract is exclusively the digital content and not the carrier as an object.[42] The differentiation practised and upheld in Germany over decades[43] has hitherto not

---

36 Cf. J. DRUSCHEL and M. OEHMICHEN (2015) *CR* 173, 176 et seq.
37 This approach is to be welcomed since it appears promising and at least leads us to expect greater clarity in the field of consumer rights in the event of defects. The reduction is also acceptable in the light of the fact that the proposal does not involve any prejudice to copyright. However, in the long term the importance of contract law for the consequences in other fields of law such as copyright should not be neglected, cf. J. DRUSCHEL and M. OEHMICHEN (2015) *CR* 173, 178 et seq.
38 Unlike the draft Regulation on the CESL (cf. *supra* n. 9), cf. Art. 107 CESL.
39 Cf. J. DRUSCHEL and M. OEHMICHEN (2015) *CR* 173, 177.
40 ECJ, *Oracle v. UsedSoft* (2011) *GRUR Int.* 1063 et seq.
41 Cf. also recital 12.
42 DRUSCHEL, *supra* n. 9, p. 355.
43 Cf. for details on the hitherto differentiating legal situation in Germany, DRUSCHEL, *supra* n. 9, p. 10 et seq.

been very convincing[44] and also does not appear to be worth pursuing from the European point of view.

3.3.3. Exceptions

Despite the broad approach, Article 3(4), (5) and (6) of the draft Directive exclude specific constellations from the scope of application. With respect to contracts without counter-performance in money, Article 3(4) lays down that the Directive should only apply where the consumer's counter-performance is in the form of *personal* data[45] that is not strictly necessary for the performance of the contract or for meeting legal requirements or is not processed by the supplier for other purposes. The same applies to *other* data that the supplier requests in order to ensure that the digital content is in conformity with the contract or meets the legal requirements, or to data is not used by the supplier for any commercial purposes. This emphasises the character as counter-performance that justifies the supplier's duties and the consumer rights proposed if the latter are infringed.[46] Many free offers concerning which the operating system on which the digital content is to be used must be communicated (e.g. when downloading Adobe Flash Player) are therefore not covered by this Directive. In contrast, free offers in which the digital content is made available free of charge for advertising purposes but in return for details such as the email address will be covered.[47]

Article 3(5)(a) of the Directive states that the Directive does not apply in cases in which the digital format is only used as a carrier for services in which the dominant element is the supplier's intellectual creative achievement. Such cases might be where expert opinions (e.g. by an architect or a lawyer) are made accessible by email or as a PDF as an attachment to an email or in the cloud. However, this does not, as might first be assumed, include cases where digital content is created. Admittedly, in the German opinion such content is often subject to the law on contracts for work and materials, but this falls within the Directive since this digital content is developed according to consumer specifications (Art. 3(2)).

3.4. RELATIONSHIP BETWEEN THE DIRECTIVE AND OTHER REGULATIONS

The draft also addresses the relationship with existing and other regulatory material. Firstly, Article 3(7) gives priority to other provisions of Union law

---

[44] Thus J. DRUSCHEL and M. OEHMICHEN (2015) *CR* 233, 233 et seq.
[45] Cf. recital 14.
[46] Cf. also recital 13.
[47] Cf. DRUSCHEL, *supra* n. 9, p. 51 et seq.

that govern a specific sector or subject matter. Secondly, the Directive is not intended to prejudice data protection (Art. 2(8)) and copyright (cf. p. 4 of the introductory explanations).[48] In addition, the Directive only affects national law to the extent that the latter itself contains rules. Article 3(9) of the Directive lists as an example rules on the formation of contracts and the validity of contracts and states that these remain unaffected.

3.5. SUPPLIER'S OBLIGATIONS

3.5.1. Supply of Digital Content

Article 5 of the Directive obliges the supplier to supply the digital content owed by contract to the consumer (Art. 5(1)(a)) or a third party which has been chosen by the consumer and which operates a physical or virtual facility making the digital content available to the consumer or allowing the consumer to access it (Art. 5(1)(b)). According to Article 5(2), the digital content shall be supplied immediately after the conclusion of the contract unless the parties have agreed otherwise (in other words, the time of performance). In cases of doubt, where supply can be to the consumer or a third party, the earlier of the two shall apply.[49]

If this provision is read in conjunction with the definition in Article 2(1) and (10) of the Directive, the supplier is required to make the digital content accessible to the consumer or a designated third party, or grant access to it. This can be in the classical form of permitting a download or sending a DVD or CD-ROM, or by providing access to content in the cloud with the possibility of streaming the content, or by simply providing cloud space as memory or access to a social network. This certainly constitutes a consumer-friendly innovation, since it is disputed whether the supplier permitting a download is in itself sufficient.[50]

3.5.2. Freedom from Defects

If agreed by contract,[51] or if the subject of pre-contractual information, and to the extent relevant for the contract, the supplier shall ensure that the digital content is in conformity with the contract (in other words, the objective concept of defect).[52] In this respect, Article 6(1) lays down criteria against which

---

[48] Cf. also recital 21 et seq.
[49] Cf. also recital 23.
[50] Supported in e.g. St. LORENZ (2012) 212 *AcP* 702, 720; rejected in e.g. DRUSCHEL, *supra* n. 9, p. 77.
[51] Cf. recital 24.
[52] Recital 25.

conformity with the contract is to be measured: quality, durability, version, functionality, interoperability, necessary features such as accessibility, continuity and security (Art. 6(1)(a)). In addition, the content must be fit for any particular purpose for which the consumer requires it and made known to the supplier (Art. 6(1)(b)).

If the contract fails to lay down clearly and comprehensively the criteria listed in Article 6(1), the digital content must be suitable for any purpose for which comparable digital content can normally be used, including functionality, interoperability and other performance features such as accessibility, continuity and security (Art. 6(2)). However, the following criteria must be taken into account: the type of the consumer's counter-performance (Art. 6(2)(a)); where relevant, international technical standards or, in the absence of such, applicable industrial codes of conduct and good practices (Art. 6(2)(b)); any public statements made by the supplier or other persons in the chain of transactions unless the supplier can show that he was not, and could not reasonably have been, aware of the statement (Art. 6(2)(c)(i)); that the statement had been corrected by the time of conclusion of the contract (Art. 6(2)(c)(ii)); or that the consumer's decision to acquire the digital content could not have been influenced by the statement (Art. 6(2)(c)(iii)).[53]

Where the digital content is to be supplied over a period of time, it shall be in conformity with the contract throughout the duration of the period (Art. 6(3)). This corresponds with the classification of such constellations in Germany, according to which they are subject to the provisions of rental/hire law.[54]

Finally, the supplier must ensure that the digital content also meets the requirements of Articles 7 and 8 of the Directive. This means that if the digital content is incorrectly integrated into the consumer's technical infrastructure, this fact shall also be regarded as a defect if the digital content was integrated by the supplier or under the supplier's responsibility (Art. 7(a)), or if the digital content was intended to be integrated by the consumer and the incorrect integration was due to shortcomings in the integration instructions from the supplier within the meaning of Article 6(1)(c) or should have been supplied in accordance with Article 6(2) (Art. 7(b)). In addition, the digital content must be free of third-party rights with respect to the use in accordance with the contract, otherwise the content will suffer from a legal defect (Art. 8). Article 8 expressly mentions such third-party rights that are based on intellectual property.

3.5.3. Liability

The supplier of the digital content is liable to the consumer for any failure to supply digital content (Art. 10(a)), for any defect at the time of the supply

---

[53] Cf. also recital 26 et seq.
[54] Cf. J. DRUSCHEL and M. OEHMICHEN (2015) *CR* 173, 177.

(Art. 10(b)), and for any defect that arises during the course of the contract if the contract provides that the digital content is to be supplied over a period of time (Art. 10(c)), which is often the case for digital content.[55]

3.6. CONSUMER RIGHTS

3.6.1. Non-performance

In the case of non-performance, or to be more precise the failure to supply,[56] the digital content in breach of the duty based on Article 5 of the Directive, the consumer shall, according to Article 11, have the right to terminate the contract immediately under Article 13. Since national law remains in force unaffected unless a provision is adopted in the Directive, the contractual claim to performance naturally continues to apply until the contract is terminated by the consumer; he can therefore also demand a repeat supply.

3.6.2. Defects

In the event of inadequate performance, Article 12 lays down that the consumer shall be entitled to have the content brought into conformity with the contract free of charge, unless this is impossible, disproportionate or unlawful (Art. 12(1)). Pursuant to Article 12(1) 2nd paragraph, bringing the content into conformity with the contract shall be deemed to be disproportionate if the costs for the supplier are unreasonable. The determination of what is unreasonable shall take into account: (a) the value of the digital content if it were in conformity with the contract, and (b) the significance of the defect for attaining the purpose for which the digital content would normally be used. In terms of time, the supplier must remedy the defect within a reasonable time after having been informed by the consumer about the defect and shall do so without any significant inconvenience to the consumer, taking account of the nature and the purpose for which the digital content is required (Art. 12(2)).[57]

Otherwise, the consumer shall be entitled to reduce the price in proportion to the defect, pursuant to Article 12(4), which lays down a proportional decrease as hitherto according to national and European law,[58] or, alternatively, he may terminate the contract pursuant to Articles 12(5) and 13. Article 13(1) states that the contract can be terminated by notice to the supplier. The legal consequences are then set out in Article 13(2). In addition, Article 14 provides for the consumer's right to damages in the event of a defect.

---

[55] Recital 43.
[56] Recital 34.
[57] Cf. also recital 35 et seq.
[58] Cf. §441 (3), Civil Code, and Art. 3(2) and (5) Consumer Sales Directive.

3.6.2.1. Termination of the Contract in the Event of Defect

In the event that the digital content does not conform with the contract, the consumer is entitled to terminate the contract in four alternative situations:

Pursuant to Article 12(3)(a), the consumer is entitled to terminate the contract in the event of a defect if remedying is impossible, disproportionate or unlawful. It is not clear whether the concept of disproportionality is objective or relative. Following the ECJ decision in *Weber v. Putz*,[59] presumably only relative disproportionality is intended, i.e. cases in which the selected manner of remedy is disproportionate in comparison with others.

Pursuant to Article 12(3)(b), the consumer can terminate the contract if the remedy is not completed within the time specified in Article 12(2).

The consumer can terminate pursuant to Article 12(3)(c) if the remedy would cause him significant inconvenience.

Finally, the contract can also be terminated if the supplier has declared that it does not wish to effect the remedy or if such is clear from the circumstances (Art. 12(3)(c)).

The contract can only be terminated if the lack of conformity with the contract concerns the functionality, interoperability and other key performance features. The latter include 'accessibility, continuity and security where required by Article 6 paragraphs (1) and (2)'. However, the consumer benefits from the fact that the burden of proof concerning conformity with the contract lies with the supplier. Ultimately, the provision serves the principle of *pacta sunt servanda* and at the same time shows that where the defect is not essential, the right to a reduction in price is regarded as being sufficient protection for the consumer.

3.6.2.2. Rescission of the Contract

Article 13(2) deals with the legal consequences of the termination of the contract. The price paid is to be reimbursed to the consumer at the latest 14 days from receipt of notice of termination. Where personal data is provided as the consumer's counter-performance, the supplier shall as a matter of principle take all measures to prevent their further use. This is expected to include the deletion of personal details.

If the content was not supplied on a durable medium, the consumer must delete the content or render it otherwise unintelligible. This is progress as compared with the solution of compensating for value in the CESL.[60] It is also conceivable, and considerably simpler and more secure for the supplier,

---

[59] ECJ, *Weber/Putz* (2011) *NJW* 2269 et seq.
[60] Cf. DRUSCHEL, *supra* n. 9, p. 335 et seq.

to make use of digital rights management (DRM),[61] as expressly referred to in Article 13(3).

3.6.2.3. Price Reduction

Under the aforesaid preconditions, the consumer can also reduce the price pro rata (Art. 12(3)), with Article 12(4) laying down that the reduction in price shall be based on the difference in value between digital content in a defective condition and fault-free digital content. If, on the other hand, the digital content is supplied on a durable medium, Article 13(2)(e) lays down that the consumer shall, at the request of the supplier and at its expense,[62] return the durable medium to the supplier (Art. 13(2)(e)(i)) and delete any usable copy or render it unintelligible and refrain from using it or making it available to third parties (Art. 13(2)(e)(ii)).

3.6.2.4. Damages

The Member States have freedom in their implementation of the right to damages (cf. Art. 14(2)). What is interesting and makes sense, however, is the reference to damage to the digital environment in Art. 14(1),[63] which includes the consumer's existing hardware and software.[64] For the rest, the draft Directive lays down that the consumer shall be put in the position he would have been in if the content had been supplied correctly.

3.6.3. Other Provisions

3.6.3.1. Right to Terminate Long-Term Contracts

If the digital content is supplied for an indeterminate period or on the basis of a contract with a term exceeding 12 months, Article 16 of the draft Directive gives the consumer a right to terminate the contract. This also applies even if the supply is only on the basis of a continuing obligation. This is similar to the provision concerning the EU Telecoms Package for mobile telephony contracts.[65]

---

[61] Cf. also DRUSCHEL, *supra* n. 9, p. 343 et seq.
[62] Cf. DRUSCHEL, *supra* n. 9, p. 327 et seq.
[63] Cf. also recital 44.
[64] Cf. DRUSCHEL, *supra* n. 9, p. 309 et seq. on the similar provision in the draft Regulation on the introduction of a CESL.
[65] Cf. Art. 30(5) of Directive 2009/136/EU [2009] OJ L337/11.

3.6.3.2. Right of Redress

As already provided for in the Consumer Sales Directive, Article 17 of the draft Directive provides for the supplier's right of redress. Accordingly, a substantial change to the legal situation is not to be expected.[66]

3.6.3.3. Enforcement

As an expression of a high standard of consumer protection, the Member States are required to ensure that the enforcement of these consumer rights is guaranteed (Art. 18(1)). To this purpose, according to Article 18(2), there should be a right of action either for public bodies or consumer organisations or other professional organisations.[67] Something similar already exists in German fair competition law[68] and the law on standard terms of business.[69]

3.6.3.4. Mandatory Nature

Contractual terms to the detriment of the consumer are prohibited by Article 19. This approach is not new in this respect, since the previous consumer protection law already contains this non-disposability on the basis of corresponding EU directives.

3.7. CONCLUSION

The Commission's draft shows promise. It takes up issues that are not directly addressed or solved in the laws of the Member States. A welcome feature is the approach of providing the consumer with harmonised rights in the event of non-performance and in particular insufficient performance, irrespective of the highly disputed classification of contracts concerning the supply of digital content. Above all, it is gratifying to see the inclusion of types of counter-performance other than money and the provisions on the rescission of performance received. In contrast to the draft of the CESL, the proposal provides for the obligation to delete content and allows the use of DRM, which does more justice to practical needs than an obligation to compensate value, thus demonstrating a willingness to explore new avenues.

---

[66] Cf. also recital 47.
[67] Cf. also recital 48.
[68] Cf. §8 (3) No. 3, Act against Unfair Competition.
[69] Cf. §4 Injunctive Relief Act.

4. ONLINE TRADE IN GOODS

The draft Directive on certain aspects concerning contracts for the online and other distance sales of goods[70] is intended to improve electronic commerce in Europe.[71] In legal theory terms, it is based strongly on the Directive on certain aspects of the sale of consumer goods and associated guarantees,[72] and is also intended to supplement the Consumer Rights Directive[73] in the field of online trade. Accordingly, it is only intended to regulate distance sales transactions between a European vendor and a consumer without fundamentally affecting national contract law concerning the conclusion, validity and effect of a purchase contract and the consequences of a termination of the contract or national claims for damages. If one compares the draft Directive with the draft Common European Sales Law,[74] the draft directive regulates only a subsector of the electronic sale of goods but as an instrument of full harmonisation (cf. Art. 3 of the draft Directive); the main focus is on conformity of the goods with the contract (cf. Arts. 4, 5 and 7), correct installation (cf. Art. 6), the time for establishing conformity with the contract (cf. Art. 8) and the consumer's remedies for faults resulting from a lack of conformity with the contract (cf. Art. 9).

In accordance with the system of default in the Consumer Sales Directive,[75] the consumer, in the event of a lack of conformity with the contract – the determination of which must also take into account pre-contractual declarations, such as advertising[76] in electronic media – must first insist on a remedy pursuant to Articles 9 and 11, by demanding, at his choice and free of charge, that the goods be brought into conformity with the contract by the supplier through repair or replacement. If this should be impossible for legal or factual reasons, or cannot be done within a reasonable time, the consumer can demand a reduction of the price or the termination of the contract pursuant to Articles 12 or 13. In this respect, therefore, the draft Directive corresponds to the law on the infringement of contractual obligations in §§434 and 437 et seq. of the German Civil Code, but without any regulation of the claim to damages.

---

[70] See *supra* n. 7; WENDLAND, *supra* n. 4, 129 et seq.
[71] Cf. LEHMANN (ed.), *supra* n. 16, *passim*; G. GOUNALAKIS (ed.), *Rechtshandbuch Electronic Business*, Beck. Munich 2003, *passim*; H.-W. MORITZ and T. DREIER (ed.), *Rechts- Handbuch zum E-Commerce*, 2nd ed., Otto Schmidt, Cologne 2005, *passim*.
[72] See *supra* n. 29.
[73] See *supra* n. 8.
[74] See *supra* n. 9.
[75] Cf. Art. 3: 'Rights of the consumer.'
[76] On the liability for advertising statements cf. M. LEHMANN (2002) *DB* 1090 et seq.; F. WEILER (2002) *WM* 1784 et seq.; F. BERNREUTHER (2003) *MDR* 63 et seq.; H.C. GRIGOLEIT and C. HERRESTHAL (2003) *JZ* 233 et seq.

What is new and particularly consumer-friendly is the provision concerning time limits in Article 14, which extends the time for establishing conformity and the limitation period for a lack of conformity with the contract to two years starting with the time the consumer acquires 'physical possession' of the goods[77] or the goods are handed over to a carrier chosen by the consumer. In contrast to the six-month time limit of §476 of the German Civil Code, Article 8(3) lays down that if a lack of conformity becomes apparent within two years, it is presumed to have existed at the time when possession of the goods was acquired unless this is incompatible with the nature of the goods or the nature of the lack of conformity.

Apart from this extension of the time limits, the draft Directive contains no surprising innovations for the remote sale of goods in electronic commerce; its wider aspects correspond with the existing *acquis communautaire* and can be expected to meet with broad and rapid consent.

5. SUMMARY

These approaches to the development of a functioning digital internal market in the EU should fundamentally be welcomed and their implementation pursued rapidly. What is particularly innovative in the field of commercial law is above all the approach of classifying data and data records as digital goods and of recognising contractual counter-performance as 'payment'. This corresponds to the modern commercial approach and should accordingly also be implemented in commercial law as part of the evolution of European contract law. In none of the EU Member States does this conflict with the legal traditions hitherto applied, and this approach should therefore be rapidly implemented as a new European creation with perhaps global exemplary character.

---

[77] In the case of installation, pursuant to Art. 8(2) this period only commences upon conclusion of such work, but not later than thirty days after physical possession of the goods has been acquired.

SUPPLY OF DIGITAL CONTENT
A New Challenge for European Contract Law

Reiner SCHULZE

1. Introduction 127
2. Current Change in Contract Law 131
3. Conformity 134
   3.1. From Sales Law to the Supply of Digital Content 134
   3.2. Specific Features of Performance 136
   3.3. Integration in the Digital Environment 136
   3.4. Third Party Rights 137
   3.5. Time for Performance 137
4. Conceptual Continuity and Innovation: Further Examples 139
   4.1. Remedies 139
   4.2. Contractual Synallagma 140
   4.3. Contract Typology 141
5. Conclusion 143

1. INTRODUCTION

It has only taken a relatively short amount of time for the supply of digital content to become a key factor in many branches of business and a determinative element in relation to market activity. One merely needs to note that since 2008 one of the major world players has brought in over US $40 billion for application developers.[1] Moreover, in terms of the Internet economy in general it has been estimated that, if the Internet were a separate country, its GDP would be the fifth largest in the world, ahead of Germany.[2] The economic importance of the

---

[1] <www.apple.com/pr/library/2016/01/06Record-Breaking-Holiday-Season-for-the-App-Store.html> accessed 02.05.2016.
[2] EXPERT GROUP ON TAXATION OF THE DIGITAL ECONOMY, 'Working Paper: Digital Economy – Facts & Figures' (Digit/008/2014, 4 March 2014).

digital economy cannot be denied and thus nor can that of the digital market as the medium for spreading the fruits of the digital revolution to potential users in all aspects of business and society. On a legal level, contract law (as market law) is the primary facilitator and promoter of this transfer and therefore its own development is linked to the technological and economic shift termed the 'Digital Revolution'.

Contract practice has already made use of freedom of contract to respond extensively to these new developments (for example, in cloud computing through the development of different types of standard contracts that can be made available online for the conclusion of B2C or B2B contracts[3]). The national legislators of several European Member States have also directed their attention to this field (such as the United Kingdom with the recent Consumer Rights Act 2015, which contains specific provisions on consumers' guarantee rights in contracts for digital content[4]). At European level, the European legislator was one of the first legislators worldwide to outline the possibility of placing these new developments in a contract law framework: in 2011 the European Commission proposed a Common European Sales Law (CESL)[5] which covered the supply of digital content and placed this in the CESL's system of contact law. Since the withdrawal of this proposal (which the European Parliament had principally approved, but which was criticised by several Member States) the European Commission has now proposed a directive on certain aspects concerning contracts for the supply of digital content (hereinafter referred to as the 'Digital Directive').[6] This proposal is now at the heart of the discussions on the further development of European contract law in light of the challenges posed by digital content. As part of the 'Digital Single Market Strategy'[7] the publication in December 2015 of the proposed Directive coincided with an additional proposal for a directive on certain aspects concerning contracts for the online and other distance sales of goods.[8] Both the rapid technological

---

[3] On the similarities and differences between these standard contracts F. BOEHM, Cloud-Computing Verträge [2016] *Zeitschrift für europäisches Privatrecht* 358.

[4] See D. BARRY et al., *Blackstone's Guide to the Consumer Rights Act*, Oxford University Press, Oxford 2016. Contracts on the supply of digital content have also been included in the Dutch Civil Code (Art. 7.5(5) *Burgerlijk Wetboek*).

[5] Proposal for a Regulation of the European Parliament and of the Council on a Common European Sales Law, COM(2011) 635 final; see R. SCHULZE (ed.), *Common European Sales Law – Commentary*, Nomos, Baden-Baden 2012.

[6] Proposal for a Directive of the European Parliament and of the Council on certain aspects concerning contracts for the supply of digital content, COM(2015) 634 final. For critical comments see C. WENDEHORST, 'Consumer Contracts and the Internet of Things' in R. SCHULZE and D. STAUDENMAYER, *Digital Revolution: Challenges for Contract Law in Practice*, Nomos, Baden-Baden 2016, pp. 189–223.

[7] Communication from the Commission to the European Parliament, the Council, the European Economic and Social Committee and the Committee of the Regions: A Digital Single Market Strategy for Europe, COM(2015) 192 final.

[8] Proposal for a Directive of the European Parliament and of the Council on certain aspects concerning contracts for the online and other distance sales of goods, COM(2015) 635 final.

developments as well as Member States' legislation in this area place the European legislator under pressure to execute a timely response. Yet from experience it would be difficult to achieve full harmonisation (as foreseen by the proposal) once widely varying rules have already arisen amongst the differing legal traditions and legal systems.[9] An extraordinarily intensive discussion of the difficult legal issues will therefore be necessary over the coming period in order to allow for a relatively swift legislative process at EU level and to avoid a fragmentation of the law for the supply of digital content in the internal market.

The necessary modernisation of contract law is, however, embedded in the adjustment of further legal issues to the changes brought by the 'Digital Revolution' – this of course does not make the task of the European legislator any easier. For instance, contracts for the supply of digital content and for digital services are often closely linked to licence agreements and therefore to copyright law (or more generally with intellectual property law). From the customer's perspective the supply of the digital content is only worth the contractually agreed counter-performance when the user rights are transferred under the licence. At this time it is however an entirely open question as to how the legal concepts of 'property' of data or 'digital content' will develop at European level and whether a harmonisation or approximation of Member State laws is appropriate and possible.

The customer may also attach equal importance to the responsibility of third parties outside of the contractual relationship with the supplier. Often, a small business will sell digital content (such as a game or other application for a tablet computer), yet the (previous or continuous) supply of this content or the platform for updates lies with a much larger company, with whom the customer need not necessarily have a contractual relationship (and is often only linked to the supplier via a long distribution chain). It is therefore necessary to extend the contractual liability in the relationship between the user (i.e. final customer) and the supplier with rules on responsibilities, in particular for the developer but also for the links in the chain between the developer and the final supplier. In this area of law one is above all faced with the development of tort law and product liability (including a possible further development with relation to quality of data[10]). However, one will also have to consider instruments of national law, such as the *action directe*. Moreover, in light of the new developments in

---

[9] On the problem of full harmonisation in relation to the failed proposal for the Consumer Rights Directive (COM(2011) 634 final) see R. SCHULZE and G. HOWELLS, *Modernising and Harmonising Consumer Contract Law*, Sellier, Munich 2009. With respect to the specific problem of the control of unfair terms and guarantee rights see S. WHITTAKER, 'Unfair Contract Terms and Consumer Guarantees: the Proposal for Directive on Consumer Rights and the Significance of 'Full Harmonisation' (2009) 5 *ERCL* 223–247.

[10] B. KOCH, 'Product Liability for Information in Europe?' in J. POTGIETER, J. KNOBEL and R. JANSEN (eds.), *Essays in Honour of / Huldigingsbundel vir Johann Neethling*, LexisNexis, Durban 2015, pp. 245–257; D. FAIRGRIEVE et al., 'Product Liability Directive' in P. MACHNIKOWSKI (ed.), *European Product Liability*, Intersentia, Cambridge 2016.

this area, the discussion cannot focus solely on the protection of the consumer or final customer. The protection of such parties through contract terms on the supply of the digital content needs to be extended by further rules on the responsibility of the developer and other third parties *vis-à-vis* the final supplier. Such rules are necessary as the entire burden of customer protection would otherwise fall on the final supplier and not the party responsible for developing the digital content and exercising extensive control over their supply. Furthermore, contract law is ultimately to be viewed in the context of other areas of law (such as data protection[11] and the public law regulation of infrastructures of the modern information society).

The European legislator is therefore faced by a most extensive task. However, it can probably only take individual steps (and probably only small steps in this time of widespread euro-scepticism). The planned Directive on the supply of digital content is one of the first such steps. It is highly likely over the course of the legislative process that both revisions and additions will be needed. The following chapter will focus on an aspect of digital challenges for European law that forms the subject matter of the planned Directive. The core will be formed by the question whether (and in any case how) the new legal answers in this field can contribute to the development of a more coherent European contract law. This particular goal was already outlined in 2003 in the European Commission's 'Action Plan'.[12] However, after the European Commission withdrew the most important legislative project to reach this objective, i.e. the Common European Sales Law, it remains an open question whether a more coherent European contract law can be reached by other means, or whether the European legislation in this area will resort to the often-bemoaned 'fragmentation' resulting in the absence of a system and even contradictions.[13]

It is therefore to be considered whether the overarching context of a European *contract law* can be guaranteed in comparison to dissolving these issues into 'sector-specific' rules pertaining to separate political objectives (consumer protection, protection of SMEs, etc.). Particular attention will also have to be paid to the role played by the new challenges brought by digital content for

---

[11] See now Regulation (EU) 2016/679 of the European Parliament and of the Council of 27 April 2016 on the protection of natural persons with regard to the processing of personal data and on the free movement of such data, and repealing Directive 95/46/EC (General Data Protection Regulation) [2016] OJ L119/1. For an overview of key aspects see C. CUIJPERS, N. PURTOVA and E. KOSTA, 'Data Protection Reform and the Internet: The Draft Data Protection Regulation' in A. SAVIN (ed.), *Research Handbook on EU Internet Law*, Edward Elgar, Cheltenham 2014, pp. 543–568.

[12] European Commission, *A more coherent European contract law – An action plan*, COM(2003) 68 final.

[13] On this and other options see H. SCHULTE-NÖLKE, 'How to realise the "Blue Button"? – Reflections on an optional instrument in the area of contract law' in R. SCHULZE and H. SCHULTE-NÖLKE (eds.), *European Private Law – Current Status and Perspectives*, Sellier, Munich 2011, p. 94 et seq. See also H. EIDENMÜLLER et al., 'Der Gemeinsame Referenzrahmen für das Europäische Privatrecht' [2008] *JZ* 529–530.

contract law. An analysis is therefore necessary in relation to how the concepts and principles of the present *acquis* can possibly contribute to the coherence of European contract law in the 'digital world'. Moreover, attention will also have to be directed to how these concepts can be further developed and extended by new approaches and how such innovations can have overarching significance for European contract law.[14]

Due to the extensive nature of this topic it is only possible to make express reference to some elements. It is however first necessary to give a brief introduction to the status of the development of European contract law that underpins the proposal. Attention will then turn to briefly indicating several core questions that may need to be looked at in a new light in order to accommodate the supply of digital content: the concept of conformity, followed by several continuations and innovations concerning remedies, contractual synallagma and contract typology.

2. CURRENT CHANGE IN CONTRACT LAW

Where the status of contract law development is concerned it is first necessary to highlight a tendency that has emerged since the 1980s: the internationalisation that has been expressed, *inter alia*, in the cross-border application of concepts and contract models (often from Anglo-American contract practice[15]) and with the growing influence of the CISG on the legislation in individual countries and world regions. The characteristic elements of this international sales law and its influence on modern contract law beyond its scope of application include, for example, the focus on contractual consensus without additional requirements (such as *cause* in present French law[16]), the concept of conformity and non-conformity as the basis for determining contractual duties and remedies, and refraining from a fault-based system of liability in favour of objective liability. Following the signing of the CISG in 1980 numerous countries – from China across Eastern Europe to the Netherlands, Germany and now also France[17] –

---

[14] For suggestions on improving the proposed Digital Directive see the contributions to this volume and, for example, a forthcoming statement of the European Law Institute working group on the CESL (of which the author is a member).

[15] R. SCHULZE and A. JANSSEN, 'Legal Cultures and *Legal Transplants* in Germany' (2011) 19 *ERPL (European Review of Private Law)* 225.

[16] The notion of cause is presently contained in Art. 1131 *Code Civil*. From 1 October 2016, once the reform of the French law of obligations enters into force, cause will no longer be required for the conclusion of a valid contract, see *Ordonnance n° 2016-131 du 10 février 2016 portant réforme du droit des contrats, du régime général et de la preuve des obligations*. On the issue of cause in the reform process see G. WICKER, 'La réforme du droit français du contrat: de la cause á la causalité juridique' in G. MÄSCH, D. MAZEAUD and R. SCHULZE (eds.), *Nouveaux défis du droit des contract en France et en Europe*, Sellier, Munich 2009, pp. 53–80.

[17] For details regarding the reform process see H. BOUCARD, 'The curious process reforming France's law of obligations' (2015) 1 *Montesquieu Law Review* 1.

have modernised their contract law or law of obligations and have taken, in one way or another, this international sales law into account in their legislative process.[18] In this respect, this international sales law marks the start of a new stage in the development of contract law.[19] For EU contract law, the Consumer Sales Directive[20] features elements of this model.[21] The Principles of European Contract Law (PECL) from the Lando Commission[22] are also based heavily on the CISG model and use it as a draft for a *general* contract law. In contrast, the UNIDROIT Principles[23] use the CISG model in order to propose, in the global framework, a law for commercial law – despite the background at national level (for instance in Spain and Germany), it is an approach that has until now received surprisingly little attention with regard to the European internal market. Internationalisation and particularly the 'success story'[24] of the CISG therefore reflect in various forms deep changes that can be described at political and economic level as 'globalisation' and 'European integration'.

However, the development of European contract law from 1990 to 2010 soon extended far beyond the content of the CISG. In contrast to the CISG, EU contract law particularly included consumer contracts in legal assimilation or harmonisation, developed criteria for the control of non-negotiated terms in respect of modern mass contracting, and included pre-contractual statements and duties in the context of contract formation.[25] Furthermore, sets of rules such as the PECL, the academic Draft Common Frame of Reference (DCFR)[26]

---

[18] Summarised in F. FERRARI, 'The CISG and its Impact on National Legal Systems – General Report' in F. FERRARI (ed.), *The CISG and its Impact on National Legal Systems*, Sellier, Munich 2008, pp. 471–478.

[19] R. SCHULZE, 'The New Shape of European Contract Law' (2015) *EuCML (Journal of European Consumer and Market Law)* 139, 144.

[20] Directive 1999/44/EC of the European Parliament and of the Council of 25 May 1999 on certain aspects of the sale of consumer goods and associated guarantees [1999] OJ L171/12 ('Consumer Sales Directive').

[21] See D. STAUDENMAYER, 'The Directive on the Sale of Consumer Goods and Associated Guarantees – A Milestone in the European Consumer and Private Law' (2000) 8 *ERPL* 547–564. As a 'source of inspiration' for European contract law U. SCHROETER, 'Der digitale Binnenmarkt für Europa und das UN Kaufrecht' (2016) 115 *Zeitschrift für Vergleichende Rechtswissenschaft* 270, 283.

[22] O. LANDO and H. BEALE (eds.), *Principles of European Contract Law. Parts I and II*, Kluwer, The Hague 1999.

[23] UNIDROIT Principles for International Commercial Contracts (2010) <www.unidroit.org/english/principles/contracts/principles2010/integralversionprinciples2010-e.pdf> accessed 18.05.2016; a revised version is expected over the course of 2016.

[24] I. SCHWENZER and P. HACHEM, 'The CISG – A Story of Worldwide Success' in J. KLEINEMANN (ed.), *CISG Part II Conference*, Iustus Förlag, Uppsala 2009, p. 125.

[25] SCHULZE, supra n. 19, p. 9.

[26] C. VON BAR, E. CLIVE and H. SCHULTE-NÖLKE (eds.), *Principles, Definitions and Model Rules of European Private Law. Draft Common Frame of Reference (DCFR Outline Edition)*, Sellier, Munich 2009; see C. VON BAR, Structure and Coverage of the Academic Common Frame of Reference' (2007) 3 *ERCL* 350–361; R. SCHULZE (ed.), *Common Frame of Reference and Existing EC Contract Law*, 2nd ed., Sellier, Munich 2009.

and the European Commission's proposal for a Common European Sales Law included matters such as avoidance due to error, fraud or unfair exploitation and the restitution following avoidance or termination.[27] The EU legislator also sowed its seeds in the freshly ploughed field of e-commerce,[28] an area of law entirely unknown during the preparation of the CISG and one which could only be considered later by individual acts of uniform law and supplementary soft law.[29] Where the supply of digital content is concerned, the European Union made advancements in 2011 with its aforementioned legislative project for a Common European Sales Law. On a European scale this CESL would have codified European sales law for the first time; on a worldwide scale it represented one of the first proposals that included digital content in all parts of an overarching system of contract law. Although the Commission has withdrawn this proposal it will nonetheless serve as an important basis for the further development of European contract law, as is shown, for example, by the references to it and the amendments by the European Parliament in the new proposals.[30] Independently thereof, the inclusion of digital content in the withdrawn proposal and the subsequent legislative initiatives on this matter at national and European levels confirm that the technological and economic developments over recent years have allowed the supply of digital content to become a new challenge that is not sufficiently covered by present national and international contract law. The European Commission has raised a series of further topics, including Internet platforms and the wide range of tasks in relation to the free flow of data in the internal market.[31]

The question can therefore be asked whether there is now a second phase of the development of contract law due to the 'Digital Revolution' despite the relatively recent start to a worldwide review of contract law as influenced by the CISG. This second phase could be more greatly characterised by international, supranational or at least mutually influential national responses to the new challenges. Digital content, just like modern, Internet-based modes of communication, can hardly be suitably regulated in the national framework alone. The question could rather be posed whether new concepts and rules

---

[27] Arts. 48–51, 172–177 CESL.
[28] Especially since the Directive 2000/31/EC of the European Parliament and of the Council of 8 June 2000 on certain legal aspects of information society services, in particular electronic commerce, in the Internal Market ('Directive on electronic commerce') [2000] OJ L178/1.
[29] United Nations Convention on the Use of Electronic Communications in International Contracts 2005, which entered into force on 1 November 2013. The full text of this Convention and the Explanatory note by the UNCITRAL secretariat are available online at <www.uncitral.org/pdf/english/texts/electcom/06-57452_Ebook.pdf> accessed 18.05.2016. Already much earlier UNCITRAL Model Law on Electronic Commerce (1996). On the capability of the CISG in relation to e-commerce see SCHROETER, *supra* n. 21, p. 286.
[30] COM(2015) 634 final, *supra* n. 5, p. 2; COM(2015) 635 final, *supra* n. 7, p. 2. See also M. WENDLAND, 'GEK 2.0? Ein europäischer Rechtsrahmen für den Digitalen Binnenmarkt' (2016) 1 *GPR (European Union Private Law Review)* 8, 18.
[31] For example cloud computing raised in COM(2015) 192 final, *supra* n. 6, p. 14.

for the contract law of the 'digital age' are actually required or whether the adequate application of existing rules would not suffice. Over the past decades, the discussions on electronic conclusion of contract gave good reasons for allowing this modern development to fall within traditional concepts and doctrine for the formation of contract.[32] In light of the expanding digitalisation of contracting – from the preparation to performance to 'unwinding' the contract – and especially due to the supply of digital content, it can instead now be asked whether this view can still be maintained without reservation or whether it will have to be reconsidered. Taking contract formation as an example, it will have to be examined whether the views advanced twenty years ago for contracting via the internet can continue without change to 'automated' conclusion of contract in the 'Internet of Things'.[33] However, these and a number of other questions cannot be considered here in detail; the following will rather have to be limited to the aforementioned examples of conformity, remedies and contract types.

3. CONFORMITY

3.1. FROM SALES LAW TO THE SUPPLY OF DIGITAL CONTENT

The notion of conformity has become one of the flexible, key concepts of modern contract law. The concept not only followed a path from the CISG into national laws but also took a route into European contract law via the Consumer Sales Directive and the DCFR.[34] The CESL expressly included

---

[32] For instance, R. BRADGATE and G. HOWELLS, 'When Surfers Start to Shop: Internet Commerce and Contract Law' (1999) 19 *Legal Studies* 287, 314, noting that a revolution of legal principle would be an overstatement. See also now in English law *Golden Ocean Group Ltd v. Salgaocar Mining Industries PVT Ltd & Anor* [2012] EWCA Civ 265 and the application of traditional principles of contract formation and authentication to e-mail correspondence. Cf. for example, A. WIEBE, 'Vertragsschluss bei Online-Auktionen' (2000) *Multimedia und Recht (MMR)* 323, 329, concluding a general need for an interest oriented modification and development of general contract law in light of new methods of communication.

[33] See, for example, the earlier work W. SUSAT and G STOLZENBURG, 'Gedanken zur Automation' (1957) *MDR* 146–147, in which a computer generated statement of intention is questioned as a genuine statement of intention for the purposes of contract conclusion.

[34] In particular, Art. IV.A.2.-301 et seq. DCFR; earlier in the Principles of the Existing EC Contract Law (Acquis Principles), especially Art. 7:B-01 et seq., 8:B-03 et seq. ACQP; printed in RESEARCH GROUP ON THE EXISTING EC PRIVATE LAW (ACQUIS GROUP), *Contract II – General Provisions, Delivery of Goods, Package Travel and Payment Services*, Sellier, Munich 2009 and also in H. BEALE et al. (eds.), *Fundamental Texts on European Private Law*, 2nd ed., Hart, Oxford forthcoming; on the methodology and function of the Acquis Principles see R. SCHULZE, 'European Private Law and Existing EC Law' (2005) 13 *ERPL* 1 and G. DANNEMANN, 'Consolidating EC Contract Law: An Introduction to the Work of the Acquis Group' in *Contract II*, pp. xxxvi–xlix.

conformity in contracts for the supply of digital services,[35] but with a link of its subjective and objective elements differing from the Consumer Sales Directive.[36] The concept of conformity now also forms a basis in the Digital Directive for the liability of the supplier and the consumer's remedies (Articles 10 and 12 Digital Directive).

For the future development of European contract law it will, however, be of considerable importance to determine which modifications to this sales law-based concept may be necessary in its application to contracts for the supply of digital content. It will be shown that there will have to be extensive changes in order to take sufficient account of the differences between the supply of digital content and the delivery of goods. Whereas such features were hardly expressed in the CESL, the proposed Digital Directive pays greater attention (although not entirely suitable in every aspect) to these features. Four aspects should be noted here: specific 'performance features' mentioned in the Digital Directive, integration into the digital environment, rights of third parties, and time aspects regarding performance. It is above all the Directive's conceptual approaches regarding the development of conformity in relation to digital content which is at the fore. The scope of this contribution does not allow for detailed discussion of the individual aspects in the proposed Directive which, despite their innovation, do feature considerable flaws and require improvement.[37] One introductory example will suffice here: like the Consumer Sales Directive, the proposed Directive combines subjective and objective elements of conformity. However, where the latter element is concerned, it is not sufficiently clear from the wording whether it simply concerns criteria which are subsidiary to a contrasting provision in the contract (i.e. non-mandatory law that would not, in addition to the Unfair Terms Directive,[38] protect the consumer against the stronger supplier)[39] or whether it concerns a minimum standard which limits freedom of contract with a mandatory aspect. In relation to this and other questions (e.g. the consequences of the consumer's knowledge of the non-conformity before the conclusion of contract) further explanation is necessary.

---

35   Especially in Arts. 91(c) and 99 et seq. CESL.
36   M. SCHMIDT-KESSEL et al., 'Die Richtlinienvorschläge der Kommission zu Digitalen Inhalten und Online-Handel – Teil 2' (2016) 2 *GPR* 54, 65; G. SPINDLER, 'Verträge über digitale Inhalte – Haftung, Gewährleistung und Portabilität. Vorschlag der EU-Kommission zu einer Richtlinie über Verträge zur Bereitstellung digitaler Inhalte' (2016) *MMR* 151. See also WENDLAND, *supra* n. 30, p. 15.
37   For further references see the ELI statement, *supra* n. 14.
38   Council Directive 93/13/EEC of 5 April 1993 on unfair terms in consumer contracts [1993] OJ L95/29.
39   See however WENDLAND, *supra* n. 30, p. 15.

3.2. SPECIFIC FEATURES OF PERFORMANCE

On the one hand, the concept of specific 'features of performance' places the supply of digital content in the general terminology of contract law: it concerns a 'performance'. On the other hand, the concept indicates that the performance is characterised by particular 'features'. In this respect, it refers to specifics that in general need not be present in contractual performance but which describe the supplier's performance in the supply of digital content (and can possibly serve to define more closely the 'typical' contractual performance in the supply of digital content). The proposed Digital Directive uses the term 'performance features' in the first subsection of its provisions on the conformity of digital content (Art. 6(1)(a) Digital Directive). According to the provision, digital content is in conformity with the content when it (as far as is relevant) corresponds to a series of performance features, including any pre-contractual information which forms an integral part of the contract. The first part of the list of features covers quantity, quality, duration, version, functionality, and interoperability. The second part covers additional 'other performance features' in, quite rightly, a non-exhaustive catalogue of criteria. Express reference is however made to accessibility, continuity and security. Even though the reasons and criteria for the split into these two groups (the 'features' and the 'other' features) may be doubtful, the list does show the intention to link general requirements typically used in sales law (such as quantity and quality[40]) with features that have particular relevance to digital content. These latter features can also refer to other contractual objects than the supply of digital content (for example, 'duration' and 'version' could also describe tangible goods in a traditional sales contract); however, they are emphasised due to their relevance for digital content. In turn, some features concerned are specifically formulated in relation to digital content (such as accessibility and interoperability). On the whole, this list of performance features gives the concept of conformity a *specific* form that clearly distinguishes its application to digital content from the application in, for instance, the Consumer Sales Directive to tangible goods[41] or to 'goods'[42] in the scope of the CESL. In this latter instance, the concept in the Digital Directive extends far beyond the (general) criteria for conformity of digital content[43] as proposed in the CESL.

3.3. INTEGRATION IN THE DIGITAL ENVIRONMENT

The same may also be said for the integration of digital content in the customer's 'digital environment'. For digital content there is the task to develop an approach

---

[40] See, for example, Art. 2(1)(d) Consumer Sales Directive.
[41] See Art. 1(1)(b) Consumer Sales Directive.
[42] Art. 2(h) CESL.
[43] See Art. 99 et seq. CESL.

specific to digital content which builds on the basis set by the Consumer Sales Directive for traditional sale of goods: defects arising from incorrect installation by the seller or due to defective instructions are deemed equivalent to non-conformity (see Art. 2(5) Consumer Sales Directive).[44] Article 7 Digital Directive transfers this approach to the common situation in which the customer can only use the digital content when it is introduced into his existing digital environment (e.g. software and operating system). Such surroundings include the customer's hardware and software environment.[45] This continuation of an approach from the present consumer *acquis* to the new requirements for protection in the digital world clearly appears to be convincing and, at least in this regard, illustrates that continuity and innovation can be linked in adjusting European contract law to particular digital challenges.

3.4. THIRD PARTY RIGHTS

A similar assessment may also apply to the principle that digital content must be free from third party rights. Article 102 CESL already extended the underlying principle in sales law to digital content (and specified this with a reference to rights or claims based on intellectual property). With its express reference to intellectual property the proposal for a Digital Directive picks up this principle as an element of conformity of digital content and (fortunately) also considers this point in relation to the supply over a period of time (Arts. 6(5) and 8 Digital Directive). Nevertheless, further additions concerning specific matters of the supply of digital content are lacking, such as clarification that End User Licence Agreements (EULA) do not affect the consumer's legal position as determined by the Directive with regards to conformity.[46]

3.5. TIME FOR PERFORMANCE

A crucial difference to the notion of conformity in sales law does however concern time aspects surrounding performance. This can be demonstrated using three examples.

Firstly, digital content can not only be supplied in a single performance but also as continuous performance over a period of time. In the latter instance, the requirement of conformity does not refer to a *point* in time but rather a *period* of time. Here it is possible to see the difference to those *sales contracts* that do not concern the single exchange of performance but rather provide for

---

[44] Also on this point SPINDLER, *supra* n. 36, p. 219.
[45] See recital 30 Digital Directive.
[46] For other issues concerning EULAs see WENDEHORST, *supra* n. 6, pp. 197–198.

repeated performance (such as the monthly delivery of goods). In such sales contracts there are numerous points in time for conformity, but a *period of time* is not decisive. The supply of digital content over a period of time rather reflects the typical type of performance in a service contract and indicates the proximity between the supply of digital content to this kind of contract – and shows the transfer of the concept of non-conformity beyond sales law (following the model adopted by national legal systems that have adopted the concept of conformity in general contract law and therefore apply the concept to all types of contracts, including service contracts as, for example, in German law in §323 BGB).

Without answering the question of the possibility of such a broad generalisation, the Commission's proposal for a Digital Directive has decided in favour of also using the concept of conformity in the supply of digital content over a period of time. In these cases, the proposed Directive therefore provides that the digital content shall be in conformity with the contract 'throughout the duration of that period' (Art. 6(3) Digital Directive). Accordingly, the supplier's liability for any lack of conformity is linked to two alternative time requirements: on the one hand the time at which the digital content is supplied; on the other hand for the duration of a period when the supply is over a period of time (Art. 10(b) and (c) Digital Directive).

Consequently, the traditional sales law rules on transfer of risk (in principle aimed at the moment of delivery to the buyer[47]) and on the burden of proof have to be modified in accordance with these requirements for the supply of digital content. Accordingly, this has a number of different consequences, though one is especially notable. As under the Consumer Sales Directive, a presumption of non-conformity also applies to the supply of digital content (but not for six months). According to Article 9 Digital Directive the burden of proof regarding the conformity is on the supplier. In this respect, the difference between the point of supply and a period of time for supply (in the relevant contracts) is also relevant (as is shown by the cross-reference in Art. 9 to Art. 10 Digital Directive).

Secondly, particular reference is also due the 'updating' requirement, which is often necessary to allow the customer to make long-term use of the digital content, as intended by the contract.[48] According to Article 6(1)(d) Digital Directive, the digital content is therefore only in conformity with the contract when it shall be updated, as stipulated by the contract. Although it is correct for the Directive to adopt this approach, it does pose a number of questions. For example, how to precisely determine the time frame for updating (whether similar standards are appropriate as in sales law decisions regarding the supply

---

[47] Art. 20 Directive 2011/83/EU of the European Parliament and of the Council of 25 October 2011 on consumer rights [2011] OJ L304/64; see also Art. IV.A.–5:102 DCFR, Arts. 142–143 CESL.

[48] For more detail see SPINDLER, *supra* n. 36, p. 220.

of spare parts[49]), the extent to which this subjective element of conformity[50] corresponds in some form to the objective element of conformity under Article 6(2) Digital Directive (especially under the aspect of 'continuity') when such term is missing; and whether the update is to be considered as a subsequent part of the contractually owed supply of digital content and is therefore covered by the presumption under Articles 9 and 10 Digital Directive.

Thirdly, the time elements are also included in other criteria for the conformity of digital content, especially in the 'continuity' and 'accessibility' which are not only subjective, but also form objective elements of conformity according to Article 6(1)(a) and (2)(a) Digital Directive. The express extension of these performance features underlines the particular significance of the time dimension and also requires concretisation in future doctrine and jurisprudence.

4. CONCEPTUAL CONTINUITY AND INNOVATION: FURTHER EXAMPLES

4.1. REMEDIES

The combination of new components with approaches already familiar from European sales law is not only demonstrated by the concepts of conformity and the liability for lack of conformity but can also be seen in the system of remedies. Only a few examples are needed to illustrate this. The basic structure outlined for sales law by the Consumer Sales Directive and further by the CESL appears to be transferable to digital content, as well as the remedies of bringing the content into conformity, price reduction, termination, and damages (Art. 12 et seq. Digital Directive). The tradition in EU law corresponds to the primacy of subsequent performance or bringing into conformity, established as a principle of sales law by the Consumer Sales Directive.[51] In accordance with this principle, the proposed Directive places the consumer's entitlement to have the digital content brought into conformity with the contract free of charge at the top of the list of remedies (Art. 12(1) Digital Directive). However, the proposal does

---

[49] It has long been recognised under German law that the requirement of good faith in §242 BGB places the supplier under an implied collateral contractual obligation to supply spare parts during the life expectancy of the good, see e.g. AG Munich (1970) *Neue Juristische Wochenschrift* 1852; AG Russelsheim, (2004) *Deutsches Autorecht* 280. The introduction of a legislative requirement concerning the reserve of spare parts was discussed in the United Kingdom in the 1980s and rejected, see Law Commission, *Sale and Supply of Goods*, Law Comm. No. 160, 1987, para. 3.66.

[50] See Art. 6(1) Digital Directive: 'as stipulated in the contract'.

[51] See, for example, the comments by BIANCA to Art. 3 Consumer Sales Directive in S. GRUNDMANN and M. BIANCA (eds.), *EU Sales Directive*, Intersentia, Antwerp 2002, pp. 168–169. For the primary of performance under the proposed Directive also SCHMIDT-KESSEL, *supra* n. 36, p. 66; WENDLAND, *supra* n. 30, p. 16.

take into account the differences between digital content and tangible goods as it does not, for example, adopt the modes of subsequent performance (repair and replacement) stated in Article 3(2) Consumer Sales Directive and therefore does not contain any corresponding rules on a party's choice between these two means of bringing the subject matter into conformity.

Furthermore, there is a clear effort to take account of the features of the supply of digital content in respect of termination of contract (despite a number of unclear points in relation to the requirements and consequences of this remedy[52]). From the supplier's perspective, he shall take measures in order to refrain from the use of data that he may have received from the consumer as counter-performance and in order to allow the consumer to retrieve data provided to the supplier, whereas the consumer shall delete or render unintelligible data that has been supplied (Art. 13(2) Digital Directive). Whether and how these new instruments for ending and unravelling performance can be realised in practice is a different question. In our context they are indeed to be valued as an expression of the effort to combine an established structure of European contract law with innovations.[53]

4.2. CONTRACTUAL SYNALLAGMA

Although the emphasis of this chapter is placed primarily on conformity and several consequences of non-conformity,[54] it ought to be mentioned that the supply of digital content also poses new questions for entirely different areas of

---

[52] For example, the time for exercising this remedy and regarding the notion of the 'main performance features of the digital content' in Art. 12(5) Digital Directive ('main' performance features is described using the same examples as for 'other' performance features under Art. 6(1) Digital Directive, namely accessibility, continuity and security, so that the question can be asked whether generally the performance features under Art. 6 are to be considered as main performance features under Art. 12(5). Accordingly, the further question arises whether there is no specific (high) threshold for the termination of the contract as for the other remedies, though as recital 35 of the Digital Directive indicates, a 'more than negligible' standard may apply under particular circumstances. For further problems see WENDLAND, *supra* n. 30, pp. 16–17.

[53] Similar comments also apply for the law of damages, even though the proposed Directive only covers a narrow aspect: Art. 14 Digital Directive contains the concept of the 'economic damage to the digital environment of the consumer' in order to limit the Directive's scope of regulation from damages to be compensated solely by national law. Here, the 'digital environment' as a legal concept can gain at least the same importance in practice as its use in Art. 7 Digital Directive in relation to the integration of the digital content 'into the consumer's digital environment'.

[54] The scope of this chapter does not allow for particular consideration of the remedies of price reduction (which the Digital Directive has developed beyond the basis set out in the Consumer Sales Directive and the CESL for digital content) and damages (in which urgent clarification is needed especially with regard to the 'detailed rules for the exercise' of this right according to Art. 14(2), namely whether the Member States can modify the objective liability under Art. 14(1) with fault-based requirements).

contract law and requires new focus in doctrine and legislation – not least with regard to the concept of the bilateral contract. Whether an agreement with a business on the supply of digital content is to be viewed as a *bilateral* contract can often depend on how one perceives the business' receipt of access to the other party's data. Particularly where the doctrine of consideration is concerned, some legal systems may require specific reasoning as to the extent to which this represents counter-performance. From an economic perspective, data can be of considerable importance to the supplier of digital content (for its own use or for sale to other businesses).[55] In providing data to the business the customer therefore gives something of great value to the business though without losing the data as part of its 'assets'.[56] Against this backdrop, the proposed Directive chooses a notable, but not entirely convincing middle path: Its broad notion of counter-performance covers performance in money as well as in 'the form of personal data or any other data'. However, the Directive's scope of application (and therefore the extent of its consumer protection) is limited to contracts in which the consumer has to pay a price or 'actively' provides counter-performance in the form of data (Art. 3(1) Digital Directive).[57] It may appear doubtful whether, firstly, this criterion of the 'active' consumer sufficiently secures consumer interests in relation to common access practices, particularly concerning personal data and, secondly, whether it accords on a theoretical level with the general concept of performance, *Leistung, prestation* (which can in principle also consist of a forbearance[58]). In any case, the positive aspect of the proposal is that it highlights the need for closer attention to the contours of 'counter-performance' and thus of contractual synallagma in the context of the supply of digital content.

4.3. CONTRACT TYPOLOGY

The question of the bilateral contract brings one neatly to the broader question whether the Directive contains specific indications of how to categorise contracts for the supply of digital content.[59] Caution is however required in this respect.

---

[55] See C. LANGHANKE and M. SCHMIDT-KESSEL, 'Consumer Data as Consideration' (2015) *EuCML* 218.

[56] As indicated in LANGHANKE and SCHMIDT-KESSEL, *supra* n. 55, p. 221. For further general comments on the significance of data as consideration see also WENDEHORST, *supra* n. 6, pp. 193–194.

[57] See SCHMIDT-KESSEL, *supra* n. 36, p. 60 et seq. and WENDLAND, *supra* n. 30, p. 14 indicating that the 'unnoticed' collection of data would not represent counter-performance for the purposes of Art. 3(1) Digital Directive.

[58] For national laws, see for example *Dunlop Pneumatic Tyre Co. v. Selfridge* [1915] AC 847 for English law; §241(1) German Civil Code (BGB). For French law see the comment by Y. PICOD, in Recceuil Dalloz, V° 'Obligations', n° 30 s. (2015).

[59] See SCHMIDT-KESSEL, *supra* n 36, p. 60 et seq.

The multitude of different economic and social interests, as well as purposes in this broad field, leads to a great variation in the forms of contract, which in turn can be subject to entirely different classifications depending on the relevant national law. For instance, cloud computing contracts, which are by no means uniform in type, are mostly classed in many European countries as service contracts, but also in part as contracts to produce a work or as goods contracts. In contrast, in Germany such contracts are often categorised as a lease or a loan contract.[60] One can therefore not expect a uniform classification of the contract with respect to the variety of contracts covered by the Directive.

However, it may be worth considering which legal models in existing EU law are followed by the provisions in the proposal and in doing so taking account of which role is played by concepts and principles which were originally applied to a particular, traditional contract type. As already shown in relation to conformity and remedies, sales law has played an influential role (which is hardly surprising in light of the prominence of sales law in the development of European contract law). However, the transfer of the concept of conformity and the remedial structure from sales law to the new world of digital content goes together with the enrichment and modification by innovative elements such as consideration of time dimensions and the relationship between the digital content and its environment ('interoperability', 'accessibility', 'digital environment'). For the most part, these components point in the direction of classification as a service contract (in the broader sense of the European concept of services). This means that the aforementioned concepts originally developed for sales will be transferred by the proposed Directive to the area of services, however without one being able to say which relevance this will have beyond the Directive's scope of application. For the development in the near future it is conceivable that the service-related criteria of conformity and the according structure of remedies are restricted exclusively to the supply of digital content and some other specific areas such as package travel.[61] However, over the long term where a general concept of service contract in European law is concerned, the emergence in these fields of such building blocks cannot be excluded. The provisions for digital content and package travel could therefore prove to be the first pillars of European services law that are based on the same fundamental concepts as sales law and thereby allowing *general* conceptual bases to strengthen the coherence of European contract law. In this case, the new European law for the supply of digital content would not just be of considerable importance for the digital internal market but also for the development of a more coherent

---

[60] See European Commission, *Comparative Study on cloud computing contracts – Final Report* (March 2015); BOEHM, *supra* n. 3, pp. 366–367.

[61] Directive (EU) 2015/2302 of the European Parliament and of the Council of 25 November 2015 on package travel and linked travel arrangements amending Regulation (EC) No 2006/2004 and Directive 2011/83/EU of the European Parliament and of the Council and repealing Council Directive 90/314/EEC [2015] OJ L326/1.

European contract law as a whole. This perspective may give cause to pay particular attention to the coming discussions on the proposed Directive and, where possible, to improve it with experiences from the Member States.

5. CONCLUSION

It is with such an outlook that a brief conclusion can be drawn in five sentences:

(1) The new challenges created by the supply of digital content for European contract law require an extension of existing concepts in the *acquis communautaire* through new approaches.
(2) The proposed CESL was somewhat premature as it did not build sufficiently on new approaches but was rather generally satisfied with adopting categories from sales law without making necessary changes.
(3) In comparison, the proposed Digital Directive combines concepts from sales law (especially the Consumer Sales Directive), such as conformity and the remedial structure, with newly developed approaches and rules in relation to digital content (in particular the time aspects of performance, the accessibility to tendered performance and the integration of the subject matter of performance into the recipient's environment).
(4) In contrast to the CESL, the proposed Directive gives greater consideration to service aspects in the supply of digital services and transfers, with the corresponding changes, sales law concepts to the contracts it covers.
(5) This transfer (especially regarding conformity and the remedial structure) to contracts beyond sales possibly gives rise to the chance to use the future *acquis communautaire* of the 'digital internal market' to come closer to a more coherent general contract law, as Ole Lando and the earlier pioneers of European contract law strived to achieve, though on a different basis, before the digital revolution.

REFLECTIONS ON REMEDIES FOR LACK OF CONFORMITY IN LIGHT OF THE PROPOSALS OF THE EU COMMISSION ON SUPPLY OF DIGITAL CONTENT AND ONLINE AND OTHER DISTANCE SALES OF GOODS

Geraint HOWELLS

1. Introduction 145
2. Online and Distance Contracts 147
   2.1. To Cure or Not to Cure? 147
   2.2. Consumer Sales Directive 147
   2.3. Common European Sales Law 148
   2.4. The Proposal for Online and Other Distance Sales of Goods Directive 149
       2.4.1. Hierarchy of Remedies: Repair and Replacement 149
       2.4.2. Replacement 152
       2.4.3. Repair 152
       2.4.4. Price Reduction and Termination 153
3. Proposal for Digital Content Directive 155
   3.1. Goods or Services Distinction 155
   3.2. Remedies for Lack of Conformity 157
   3.3. Damages 159
   3.4. Life-Time Contracts 160
4. Conclusions 160

1. INTRODUCTION

The EU has been seeking to reform its sales law for some time. It has not been concerned so much to amend the basic principles of the law, but rather feels the minimum harmonisation nature of the Consumer Sales Directive impedes cross-border trade.[1] Its attempts to include reform of sales law within the Consumer Rights Directive were rebuffed, because it was seeking to impose maximum harmonisation at a level that several Member States could not accept. This was principally due to its Proposal seeking to require consumers to accept cure (i.e. repair or replacement) as the first option before rejection and termination were possible.[2] Its second attempt to promote a cross-border regime for consumer sales of goods by way of an optional Common European Sale Law[3] was also withdrawn due to opposition based on the only option in practice being given to the trader, who would decide whether to trade on the basis of CESL. Also the scheme would have removed the consumer protection afforded by the Rome I Regulation rules which ensures consumers retain the protection afforded by national mandatory rules. The Commission is trying for a third time by introducing a proposal for a maximal harmonisation directive limited to online and other distance contracts. At the same time the EU is seeking to engage with the digital agenda.[4] As part of that process it has proposed a directive on sales of consumer goods online and other distance sales of goods[5] and a directive regulating conformity in the context of digital products.[6]

The basic core principles of conformity have been well settled for sale of goods. The real battle ground has been over the appropriate remedies regime. This will therefore be the focus of this chapter. There might be more to discuss about the conformity concept in relation to digital goods, but our concentration will still be on the remedies regime. The punchlines of this chapter are that the EU should accept that Member States prefer to retain their national diversity as regards remedies for non-conformity in consumer sales contracts and withdraw the online and distance sales proposal. By contrast their work to create an appropriate regime for digital contracts is to be welcomed. Few states have clear rules on the law that applies to digital products and clarification is to be welcomed. The United Kingdom has adopted a special regime for digital content in the Consumer Rights Act 2015 and its rules will be compared with those proposed by the Commission.

---

[1] Directive 1999/44/EC on certain aspects of the sale of consumer goods and associated guarantees [1999] OJ L171/12 (hereafter Sale of Goods Directive).
[2] See Proposal for a Directive on Consumer Rights, COM(2008) 614 final.
[3] Proposal for a Common European Sales Law, COM(2011) 635 final. This included a Regulation on a Common European Sales Law and the text of the Common European Sales Law.
[4] Commission Communication, *A Digital Single Market Strategy*, COM(2015) 192 final.
[5] Proposal for a Directive on certain aspects concerning contracts for the online and other distance sales of goods, COM(2015) 635 final (hereafter, Proposal for Online and Other Distance Sales of Goods Directive).
[6] Proposal for a Directive on certain aspects concerning contracts for the supply of digital content, COM(2015) 634 final (hereafter Proposal for Digital Content Directive).

2. ONLINE AND DISTANCE CONTRACTS

2.1. TO CURE OR NOT TO CURE?

The civil and common law have instinctively different responses to a breach of contract. Civil lawyers believe the debtor should be required, and also permitted, to try to remedy the non-performance and have a second attempt to perform the promised obligation. Civilian lawyers argue contract law should be about fulfilling the promises made and satisfying the expectations created by the agreement. If this is achieved, albeit not at the first attempt, who can complain? Thus the right to cure is the primary remedy with the debtor being allowed to repair the goods to bring them into conformity. The common law takes a more economics-based approach. The debtor had their opportunity to perform and seek counter-performance. If they supplied non-conforming goods they exposed themselves to the risk that the other party may not allow them a second chance. The creditor should not have to expose himself to the risk that a debtor who has shown himself unable to fulfil his obligations will be able to remedy the non-conformity. He should just be able to walk away. This applies even if his motive is entirely one of self-interest. For instance, he might have come to regret the bargain or seek to take advantage of circumstances having changed. Although there is no general rule that consumers can simply walk away from obligations if they change their minds, this is possible in the common law where the other has breached the contract by supplying non-conforming goods. Of course the practice is often different from the law in the books and (as is often the case) the differences between civil and common law are less stark in reality. Even under the common law consumers will often allow the trader to attempt to cure and the right to reject might even be preserved whilst cure is attempted. Civil law judges will not normally require cure where it is impossible or the costs disproportionate. Civilian courts will not even usually order specific performance forcing a reluctant trader to cure against his wishes and will prefer in most cases to award damages equal to the costs of cure instead.

2.2. CONSUMER SALES DIRECTIVE

The Consumer Sales Directive adopts the civilian approach. It provides for a hierarchy of remedies with the consumer first having the option to choose between repair and replacement (unless either is impossible or disproportionate).[7] The remedies of price reduction or rescission on the contract only become available if neither repair nor replacement is appropriate, or these remedies have not been completed within a reasonable time or without

---

[7] Art. 3(3).

significant inconvenience to the consumer.[8] However, conflict with the common law approach of allowing rejection immediately on breach of condition – indeed an approach that is also mirrored in some civilian countries – is avoided because the directive expressly allows Member States to maintain in place more stringent provisions in order to ensure a higher level of consumer protection.[9]

The Proposal for a Directive on Consumer Rights had proposed a similar hierarchy of remedies. In some ways the rules were even more consumer friendly. Although the trader would have been given the clear choice as to whether to repair or replace the goods,[10] on the other hand it was made clear that any remedy including rescission could be available if the same defect reappeared more than once within a short period of time[11] and that damages were available for any losses not remedied by other remedies.[12] However, its full harmonisation approach made this remedies approach unacceptable to several Member States.

2.3. COMMON EUROPEAN SALES LAW

The Common European Sales Law was the Commission's response to the failure to achieve maximal harmonisation for sales law under the Consumer Rights Directive. It sought to circumvent opposition by making it optional and limiting its scope to consumer contracts and contracts with SMEs.[13] Supporters of the Common European Sales Law pointed out that it afforded a high level of consumer protection. As regards remedies it gave consumers a free choice of specific performance (including repair or replacement), termination of the contract and return of the goods, price reduction and claiming damages.[14] The lack of practical choice on the consumer's part as to whether the contract was based on the Common European Sale Law remained a major concern. There was also a belief that traders would have to see some advantage in opting into this regime and so there was a potential risk that either the remedies would be diluted as the measure passed through the political process or would in due course be amended. Consumers preferred the security of their national mandatory law. There were also concerns about the creation of a dual regime for cross-border and domestic sales.[15] The measure lost political support and was withdrawn, with the Commission promising a more focused approach on online and digital products, which is the main subject matter of this chapter.

---

[8] Art. 4(5).
[9] Art. 8(2).
[10] Art. 26(2).
[11] Art. 26(4).
[12] Art. 27.
[13] Regulation on a Common European Sales Law, Art.7.
[14] Common European Sales Law, Art. 106(1)(3).
[15] Member States would have had the option of making it available for domestic contracts: Regulations on a Common European Sales Law, Art. 13(a).

2.4. THE PROPOSAL FOR ONLINE AND OTHER DISTANCE SALES OF GOODS DIRECTIVE

Although online is prominent in the title of this Directive the provisions are actually legally based around the broader concept of distance selling.[16] A distance sales contract is defined as 'any sales contract concluded under an organized distance selling scheme without the simultaneous physical presence of the seller and the consumer, with the exclusive use of one or more means of distance communication, including via internet, up to and including the time at which the contract is concluded'.[17]

Its rules on conformity as regards quality are similar to those in the Consumer Sales Directive, but follow the Consumer Rights Directive in also requiring they comply with any pre-contractual statement that forms an integral part of the contract.[18] They are broader in also covering quantity and freedom from third party rights.[19]

2.4.1. Hierarchy of Remedies: Repair and Replacement

The remedies for lack of conformity are set out in Article 9. This has a clear hierarchy of remedies. Before being qualified to access the remedies of price reduction or termination, the consumer must first claim repair or replacement. This will be free of charge. Article 11 provides the choice between repair and replacement is with the consumer unless the option chosen is impossible, unlawful, or compared to the other option disproportionate. In making this an assessment all the circumstances are to be taken into account, including the value of the goods if there was no lack of conformity, the significance of the lack of conformity and whether the alternative remedy could be completed without significant inconvenience to the consumer. This provision as drafted requires that one of the repair and replacement remedies must always be available unless impossible or unlawful.[20] As discussed below, the same approach has been approved by the CJEU on the basis of the similar rule in Article 3(3) of the Consumer Sales Directive. The excuse of disproportionate costs only applies as between repair and replacement. The fact that both those remedies impose disproportionate costs compared to the alternatives of price reduction or rescission is not relevant under the formulation set out in Article 11. Thus if replacement is impossible then repair must be attempted, unless impossible, however expensive, even if a price reduction or rescission would seem a reasonable alternative. Equally, if goods cannot be repaired they must be

---

[16] Art. 1.
[17] Art. 2(e).
[18] Art. 7.
[19] Art. 4(1)(a)(c).
[20] These are also excuses under Art. 9(3)(a).

replaced, seemingly without consideration of cost. Below cases are discussed where this was a live issue as the costs involved in remedying the defect were far higher than the costs of simply providing replacement goods due to them having been installed and the consequential costs of removing and reinstalling.

When the UK first implemented the Directive, it allowed the trader to be excused from the obligation to cure not only if the costs were disproportionate between repair and replacement, but also as between the alternative price reduction and rescission remedies.[21] The comparable remedies in the Consumer Rights Act 2015 do not include this broader comparison of costs.[22] This is in line with the Directive, but one might question whether the previous policy in fact reflected a better solution. In practice, consumers might well settle for alternative remedies in such instances and one finds it hard to believe a court would require such specific performance where this would be unreasonable and would instead impose a monetary award. Comment will be made below on the particular issue of installed goods, where it will be argued that even high repair costs in relation to the price of goods might reasonably be justified in some circumstances. In many cases though, allowing the Court to award a different remedy would be helpful, especially as under the Directive the repair and replacement remedies are available even for minor defects.

This issue has been tested in relation to installed goods under the Consumer Sales Directive in the European Court. In two cases referred by German courts there was an issue about the cost of replacing goods when they had been installed. *Weber v. Wittmer*[23] concerned the supply of polished tiles with shading. Two-thirds of the supplied tiles had been laid. *Putz v. Medianess Electronics*[24] concerned a defective dishwasher installed in a kitchen. One issue was whether the remedy or cure could be refused because it was disproportionate. The Advocate General would have allowed consideration of the disproportionality of the proposed replacement remedy where repair was impossible. The CJEU, however, considered that the test of disproportionality should be limited to a choice between the alternatives of repair and replacement. The Advocate General's approach seems to provide greater justice, though the true test would lie in the detail of its application. He would also not have allowed the consumer to recover the costs of taking out the defective goods and replacing them with conforming goods. However, there is a strong argument that although high in relation to the value of the goods, these costs may still not be disproportionate. When you supply goods that are capable of being installed, it is obvious that replacement may incur considerable costs. Normally the seller should have to bear this, as otherwise the costs do not disappear, but rather have to be borne by the innocent consumer. It may therefore not be disproportionate to impose

---

[21] s. 48B(3)(b) and (c) Sale of Goods Act 1979, repealed by Consumer Rights Act 2015.
[22] s. 23(3) Consumer Rights Act 2015.
[23] C-65/09.
[24] C-87/09.

significant costs on the seller, who can in any event seek redress against his supplier. However, there should be the flexibility to award an alternative remedy when for instance the defect is minor and the cost of replacing would be significant. The Advocate General would almost certainly not have been as generous to consumers as this proposed approach.

The CJEU was clear that disproportionality only applies between the remedies of repair and replacement. However, it then confused matters by stating that Article 3(3) does not prevent the Court limiting the amount recovered to an amount proportionate to the value the goods would have if there was no lack of conformity and the significance of the lack of conformity. This takes factors involved in assessing disproportionality of the alternative remedy and without any textual justification turns them into reasons for controlling the amount recoverable once the remedy is required to be granted. One can only conjecture that this was a compromise included to assuage a minority who agreed with the Advocate General's approach. It is even more difficult to reconcile this approach with the view that '[t]he possibility of making such a reduction cannot therefore result in the consumer's right to reimbursement of those costs being effectively rendered devoid of substance, in the event that he had installed in good faith the defective goods, in a manner consistent with their nature and purpose.'[25] In these installation scenarios the factor increasing the cost is the removal and reinstallation, which has to be placed on top of the cost of the replacement goods. Thus in practice there seems little opportunity to make any reduction if the main remedy of replacement is to be allowed. The better approach would be to make disproportionate cost a reason for preferring any of the alternative remedies, but not necessarily seeking to limit the costs of removing and installing unless, for example, this could be considered disproportionate as there was only a minor defect.

The CJEU had found that replacement incorporated the costs of removing the defective goods and reinstalling the replacement goods.[26] It did this partly on the basis of a textual reading of the Directive:

> '"replacement", it should be noted that its precise scope varies in the different language versions. While in some of those language versions, such as the Spanish ("sustitución"), English ("replacement"), French ("remplacement"), Italian ("sostituzione"), Dutch ("vervanging") and Portuguese ("substituição"), that term refers to the operation as a whole, on completion of which the goods not in conformity must actually be "replaced", thus obliging the seller to undertake all that is necessary to achieve that result, other language versions, such as in particular the German language version ("Ersatzlieferung"), might suggest a slightly narrower

---

[25] Para. 76.
[26] The Advocate General found this liability too broad given the lack of a filter such as causation, remoteness and fault and preferred to exclude consideration of what had occurred after the passing of risk.

reading. However, as the referring courts point out, even in the German language version, the term is not restricted to the mere delivery of replacement goods and could, on the contrary, indicate that there is an obligation to substitute those goods for the goods not in conformity.'[27]

It also noted the consumer protection principles in its previous court decision in *Quelle*[28] that remedies should be provided free of charge and that where the consumer is not at fault it was fair to impose the costs on the seller as they would have been avoided if the seller had correctly performed his contractual obligations.

The CJEU *Quelle* decision is reflected in the Proposal for Online and Other Distance Sales of Goods Directive. Article 10 makes it clear that the seller must take back the replaced goods at the seller's expense, unless agreed otherwise after the consumer has alerted the seller to the lack of conformity.[29] The obligation to replace includes bearing the costs of removing non-conforming goods and installing replacement goods.[30] The consumer is not liable for any use prior to replacement.[31]

2.4.2. Replacement

Any replacement should be completed within a reasonable time and without any significant inconvenience to the consumer.[32] In making this assessment account should be taken of the nature of the goods and the purpose for which the consumer required the goods.

It is not possible to replace exactly certain types of goods. Second-hand goods may be fairly standardised in certain sectors. For example, cars of a certain age, mileage and condition have established prices. But individual cars also always vary with regard to those factors. Thus one second-hand car cannot be an exact replacement for another. Equally, bespoke goods may be hard to replace. If the modifications are standardised so they can be replicated, for example accessories to cars, then exact replacement may be possible. If they involve creative additions which make them unique then it will be impossible to replace them.

2.4.3. Repair

As with replacement, any repair should be completed within a reasonable time and without any significant inconvenience to the consumer.[33] The conditions for

---

[27] Para. 54.
[28] Case C-404/06, *Quelle AG v. Bundesverband der Verbraucherzentralen und Verbraucherverbände*.
[29] Art. 10(1).
[30] Art. 10(2).
[31] Art. 10(3).
[32] Art. 9(2).
[33] Art. 9(2).

repair are not specified. There is no rule that the trader should only, for instance, have one chance to repair the goods. There is also no rule on whether the trader should have to collect goods or bear costs of goods being returned for repair.

2.4.4. Price Reduction and Termination

If repair or replacement are impossible or unlawful, have not been completed within a reasonable time, would cause significant inconvenience to the consumer, or the seller has declared (or it is equally clear from the circumstances) that the seller will not bring the goods into conformity within a reasonable time, the consumer is entitled to price reduction or termination.

The remedy of price reduction provides for a reduction proportionate to the decrease in the value of the goods received compared to their value if the goods had been in conformity with the contract.[34] This is a remedy the common law was unfamiliar with. The basic idea is clear enough. A table with a scratch or a car whose radio does not work, for instance, will have a lower value than perfect goods. But how will this be calculated. How is value (or value reduction) to be ascertained? Should this reduction be related to the actual price paid or the market price, as it is always possible that a consumer has paid too much or obtained a great bargain for goods? As the text refers to value, it should probably be the market price that is taken to ascertain the value of the goods with the non-conformity. This could then be compared to the market price without the non-conformity to establish a percentage decrease in value, which is then applied to the market price. Another approach is to take the actual value with the non-conformity and reduce this from the actual price paid. Many courts may find the second method the simplest, but perhaps the former complies more closely with the wording of the Directive. Price reduction is of most use when the price has not yet been paid and in can be useful in civilian systems where fault may be needed to claim damages. Where damages are available, it will normally be preferable to claim for them.

Article 13 contains rules on termination, which can exercised by notice given by any means.[35] If only some of the goods delivered lack conformity, termination may only be exercised in relation to those goods along with any goods acquired as an accessory to them.[36] This seems to risk potential injustice. 'Accessory' is not defined. Narrowly defined, it might include things which are directly used with a product. So a lens might be an accessory to a camera. Would volume A in a series of books be an accessory to volume B? In most cases it might well be so classed, but if one buys a sofa and chair are they accessories so both can be rejected? There may even be cases where items are clearly not accessories, but it may be

---

[34] Art. 12.
[35] Art. 13(1).
[36] Art. 13(2).

important for the consumer to have then all or not at all. A consumer might for instance wish to have a hedge made from 100 roses; if 50 have to be rejected he may legitimately prefer to reject them all. It would be better if the consumer was given the option of rejecting part or all of a non-conforming batch.

On termination the seller should reimburse the consumer the price without undue delay and in any event within 14 days of receipt of notice of termination and within the same timeframe the consumer should return the goods to the seller.[37] This shall be at the seller's expense. If the goods cannot be returned because of loss or destruction, the consumer shall pay the seller their monetary value at the date when the return was to be made.[38] Note this does not give the seller the right to retain the price. The value of the goods may even without a defect be different from the price and would be further reduced by the non-conformity.

Article 13(3)(d) is curious provision, which is also drafted in a complex manner, but essentially provides for a positive obligation that on termination consumers shall pay for any decrease in value that exceeds depreciation through regular use up to the price paid for the goods. Under the Consumer Sales Directive, the *Quelle*[39] decision had provided that the duty to provide the remedy of replacement 'free of charge' meant that a duty to compensate for use could not be imposed when the goods had been replaced. There had been no rider about use exceeding regular use. This rule has in fact more in common with the provision in the Consumer Rights Directive that makes the consumer liable when he has exercised his right of withdrawal to compensate for use resulting from the handling of the goods other than what is necessary to establish the nature, characteristics and functioning of the goods.[40] However, the situation is different here. The termination is caused by the seller failing to perform his obligations. The fact a consumer may have excessively used the goods or used them for unusual purposes should not *per se* mean he should have to compensate the seller for anything. If the non-regular use contributes to the lack of conformity or its effects that is a different matter and a remedy would be excluded under a separate provision: Article 9(5).

The remedies under the Proposal for Online and Other Distance Sales of Goods Directive are familiar to anyone who knows the Consumer Sales Directive. The difference between those two directives is that under the latest proposal there will be maximal harmonisation with Member States not being able to maintain or introduce divergent provisions, which either include less or, crucially also, more stringent rules.[41] This will have its most significant

---

[37] Art. 13(3)(a) and (b).
[38] Art. 13(3)(c).
[39] Case C-404/06, *Quelle AG v. Bundesverband der Verbraucherzentralen und Verbraucherverbände*.
[40] Art. 14(2).
[41] Art. 3.

impact by removing the automatic right to reject in countries like the United Kingdom. However, it will be noted that there is no right to damages under the Proposal and if this is also seen to be covered by maximal harmonisation, that will also be a significant reduction in rights. The Commission failed with this maximal harmonisation policy in the Consumer Rights Directive and even when it made such an approach optional under the Common European Sales Law. It is surprising that they have come back with another mandatory regime. Admittedly, it only applies to online and other distance contracts, which are the cross-border contracts the Commission is keen to promote. But it applies to all such contracts even if there is no cross-border dimension. It is likely to provoke similar opposition from consumer groups as the preceding proposals have. The Commission must be hoping to grind them down so eventually from weariness their opposition becomes less vocal and Commission officials can mark up a success. There is also the risk, however, that businesses will not welcome this. They will now be forced to have two distinct legal regimes if they operate a distance selling regime and one that is high street-based. One cannot imagine they will welcome that. It will also serve to confuse consumers about their rights.

3. PROPOSAL FOR DIGITAL CONTENT DIRECTIVE

3.1. GOODS OR SERVICES DISTINCTION

There has long been a debate as to how to categorise digital products.[42] Should they be treated as goods and subject to the non-conformity regime, or services and subject essentially to a negligence regime? Where software is incorporated into a product, such as to regulate cycles in washing machines or to help navigate a plane, there is little doubt that the seller of the final product is also liable for the defective software. This would also seem to be the case where hardware and software are supplied together. It seems invidious to distinguish between whether the software is preloaded or not. It is more complicated when the software is applied alone. There have been attempts to draw distinctions so that only software supplied on a durable medium is goods.[43] However, this is rather artificial as there is a distinction between the qualities expected of the durable medium itself, say a CD, and the digital content. Indeed this debate is sounding rather passé as increasingly it is common practice simply to download software or indeed just access it via a cloud. Another distinction sometimes drawn is between standardised goods and bespoke goods. This makes some sense if you look to the policy of regulating mass produced goods and digital products in the same way. But it is not a distinction drawn generally in the law for goods.

---

[42] A good summary is found in R. BRADGATE, *Consumer Rights in Digital Products*, BIS, 2010.
[43] *International Computers Ltd v. St. Albans District Council* [1996] 4 All ER 481.

Hand-crafted goods have normally to meet the same legal expectations as mass-produced goods. A more appealing distinction is between software which performs a function and software that simply provides the consumer with information on which to make decisions.[44] The latter looks akin to information provision, which is normally judged by negligence standards, whereas the former performs functionally similar to a product.

The policy choices have traditionally been seen as being between either applying the negligence standard or shoehorning digital products into the sale of goods regime. The latter task has usually been undertaken by the courts flailing around making arbitrary distinctions as discussed above. In New Zealand they simply provided in their Consumer Guarantees Act 1993 that to avoid doubt goods included computer software. Bradgate found no significant criticism of this solution in the literature.[45] However, it does not properly reflect the nature of much digital content, which actually is often merely a copyright licence. Another option is to view these contract as *sui generis*. The courts have sometimes characterised them as hybrid contracts – part sale, part licence.[46]

Given the uncertainty surrounding this issue, the approach taken by the United Kingdom was to create a separate regime for digital content in its Consumer Rights Act 2015.[47] The EU is to be applauded for following this approach in the Proposal for Digital Content Directive. This is valuable as the nature of digital products calls for different rights. For instance, many programmes contain bugs or these develop over time. If programmes could be rejected for every bug there would be no certainty, but equally the consumer needs to have them repaired. Also, being intangible, digital products cannot of course be physically returned. Many consumer digital products are for entertainment. Books and music need to be available in forms that can be easily accessed at the right quality. Computer games need to be playable. Other software actually helps perform functions, such as keeping a diary or book-keeping. Others even help run the computer itself or protect it from virus attacks. If they do not work there is a danger they cause data to be lost, interfere with the running of other programmes, expose the computer to the risk of virus or hacking attacks and cause damage to the hardware being used by the consumer. The consumer also often needs the digital content to be available on an ongoing basis and there are issues about how to redress the consumer if it stops being available. There are complex issues about whether digital content lacks conformity, but this chapter simply looks at the new remedies for what one might describe as breaches of inherent quality, external risks and lifetime dimensions of digital content agreements.

---

[44] K. ALHEIT, 'The applicability of the EU Product Liability Directive to software' (2001) 34 *The Comparative and International Law Journal of Southern Africa* 188.
[45] BRADGATE, *supra* n. 42, p. 57.
[46] *Beta Computers (Europe) Ltd v. Adobe Systems Ltd* 1996 SLT 604.
[47] Chapter 3.

The question has been raised[48] whether in fact placing all digital content into the same regime may even be over-simplifying the problem. As mentioned, digital content spans a broad range and they may not all be adequately addressed by one regime. However, it seems better to place them under this category where the terms and remedies can be fashioned to address the range of digital content than to force them into a less suitable category such as goods.

3.2. REMEDIES FOR LACK OF CONFORMITY

The broad structure for remedies is similar to that in the Proposal for Online and Other Distance Sales of Goods Directive. In the first instance the consumer is entitled to have the digital content brought into conformity free of charge.[49] The terms 'repair' or 'replacement' are not used, but in effect the same remedies will be achieved by sending another link to download (replacement) or patching any bugs (repair). The entitlement to bring into conformity does not exist where it is impossible, unlawful or disproportionate. Disproportionality relates to the costs imposed on the supplier being unreasonable taking into account the value of the digital content if it were in conformity with the contract and the significance of the lack of conformity for attaining the purpose for which digital content of that same description would normally be used. As this Proposal does not specify repair or replacement, there are no rules for choosing between alternative remedies. The supplier would seem to have the free choice as to how to bring this into conformity. This must be achieved within a reasonable time and without significant inconvenience to the consumer.[50] Equally, if costs are disproportionate it would seem no steps need to be taken. This contrasts with the United Kingdom's Consumer Rights Act 2015, which uses the terms 'repair' and 'replacement' and allows the consumer to choose between those two remedies, with disproportionate costs being assessed solely between those two remedies, as under the Proposal for Online and Other Distance Sales of Goods Directive.

Under the Proposal for Digital Content Directive, additional remedies of price reduction and termination become available in certain circumstances. These are where digital content cannot be brought into conformity as it is impossible, disproportionate or unlawful; or such remedy has not been provided within a reasonable time; or it would cause significant inconvenience to the consumer; or the supplier has declared, or it is clear from circumstances, that he will not bring the contract into conformity.[51]

---

[48] At seminar at Ferrara University, Italy, where this paper was first presented.
[49] Art. 12(1).
[50] Art. 12(2).
[51] Art. 12(3).

Price reduction again involves a proportionate decrease in the value of the digital content received compared to the value if it were in conformity.[52]

For goods, termination is only available for non-minor defects. For digital content it is available for any impairment of functionality, interoperability or the *main* performance features such as its accessibility, continuity and security. The supplier bears the burden of proving there is no such impairment.[53]

Termination must be by notice, but this can be given by any means.[54] On termination the price must be reimbursed without undue delay and in any event within 14 days of receiving the notice.[55] In addition the supplier must cease to make use of any counter-performance other than money and any other data collected.[56] Thus if the consumer had provided data this should be deleted, as should any content unless this had been created jointly by the consumer and others who continue to make use of the account. So if two people had been working in a document in a service such as Dropbox, it should remain available to the other. Equally, if someone had contributed to a design that others also worked on then it should remain available for the others when the consumer ends his contract. These factors are not covered by the Consumer Rights Act 2015, showing the value of different legislators addressing new forms of liability such as for digital content.

The supplier may prevent further use of the content by the consumer.[57] Content may be made inaccessible to the consumer or his user account disabled. However, the supplier must make it possible for the consumer to retrieve any content provided or any data produced or generated through the consumer's use of the digital content.[58] This should be free of charge, without significant inconvenience, in a reasonable time and in a commonly used format.

For the consumer's part, on termination he shall not be liable to pay for use made of digital content prior to termination.[59] There is an obligation to refrain from using digital content that was not supplied on a durable medium or from making it available to third parties.[60] It may be deleted or rendered otherwise unintelligible. Where supplied on durable medium it should be returned at the supplier's expense, but only if the supplier has requested him to do so. Any usable copy should be deleted or rendered unintelligible and should be in no way used or made available to third parties.[61]

---

[52] Art. 12(4).
[53] Art. 12(5).
[54] Art. 13(1).
[55] Art. 13(2)(a).
[56] Art. 13(2)(b).
[57] Art. 13(3).
[58] Art. 13(2)(c).
[59] Art. 13(4).
[60] Art. 13(2)(d).
[61] Art. 13(2)(e).

There is also a curiously worded provision where digital content has been supplied in exchange for a price over a period of time. This provides that consumers may only terminate for that part of the period during which the content was not in conformity.[62] The usual remedies apply except that the supplier can continue to use data and content provided by the consumer and payment should be reimbursed for the period when the digital content was not in conformity.[63] The policy seems reasonable – to allow reimbursement of price for distinct periods where the product was not available. However, the language of termination seems inappropriate. The contract is not terminated. The terminology in the United Kingdom's Consumer Rights Act 2015 is preferable. It talks about a right to refund (in the context of breach of the right to supply) and then limits this right where only some of the digital content is in breach to only that part affected by breach.[64] However, in some circumstances where the period represents a long portion of the contract period or there are several interruptions in service, full termination might indeed be appropriate.

3.3. DAMAGES

Non-conforming digital products may not only fail to fulfil their functions, but they may also cause additional damage. They may interfere with other programmes, damage the hardware, or introduce viruses that may cause malfunction or expose the computer to attack. Such damages are potentially covered by Article 14, which holds the supplier liable for economic damages to the digital environment caused by a lack of conformity or a failure to supply digital content. The digital environment includes hardware, digital content and any network connection to the extent they are within the control of the user. The aim is to put the consumer in a position as nearly as possible to that he would have been in if the content had been duly supplied in conformity. The phrasing 'as nearly as possible' is a little strange given the remedy is monetary. The objective should be to put him in the same position as he would have been in, save that of course he cannot be in that position and money has to be a substitute. Member States can develop their own rules, but recital 44 questions whether a discount on future supplies, especially when offered as an exclusive compensation, satisfies the requirement of putting the consumer as nearly as possible into the position he would have been in. The damages are for 'economic damage'. This is not defined, but presumably excludes product liability where the non-conformity causes physical damage to the product. Normally the economic damages are the cost of putting right the damage, for example repairing systems.

---

[62] Art. 13(5).
[63] Art. 13(6).
[64] s. 45.

However, there may also be economic losses if data is lost or corrupted. However, in the consumer context these are likely to be limited. Economic damages probably excludes damage for distress caused because one's data has been hacked into.

3.4. LIFE-TIME CONTRACTS

Many digital content contracts are lifetime in the sense that they are ongoing.[65] A service like iTunes might be used for your entire life or at least for a very long time. The Proposal wisely includes some rules to regulate this aspect. Under certain conditions the supplier may alter functionality, interoperability and other main performance features. The consumer is then given the right to terminate the contract within no less than thirty days from the receipt of the notice.[66] If no period is specified in the contract, the right of termination exists for at least thirty days. However, it would be useful to include an obligation to state a period of at least thirty days in the contract.

The consumer is also entitled to terminate contracts of indeterminate duration or that exceed 12 months once the first 12 months have expired.[67] Termination becomes effective 14 days after notice has been given. Where there is a price for digital content, the consumer remains liable for the period until termination becomes effective. Similar to the general effects of termination, the supplier should refrain from use of any counter-performance other than money provided by the consumer and other data collected by the supplier including content provided by the consumer; provide the technical means for the consumer to retrieve content and data; and prevent any further use of the digital content.[68] The consumer shall delete any usable copy of the digital content, render it unintelligible or otherwise refrain from using it. There is no obligation to return durable copies, presumably because even if digital content had originally been provided in that format the durable copy could have long since been disposed of or lost.

4. CONCLUSIONS

The Proposal for Online and Other Distance Sales of Goods Directive is not to be welcomed. It seeks to introduce a hierarchy of remedies in a limited, but

---

[65] L. NOGLER and U. REIFNER (eds.), *Life Time Contracts*, Eleven International Publishing, The Hague 2014.
[66] Art. 15(1)(c).
[67] Art. 16.
[68] Art. 16(4)(a)(b) and (5).

still broad category of consumer contracts. This was rejected in relation to the Consumer Rights Directive and the Commission should recognise that its approach is not supported by significant sections within the Union and several Member States. The EU needs to respect and leave room for national diversity. It also continues the potential risk of imposing a disproportionately expensive repair or replacement on a trader. The rules of the market will become fragmented depending on the means of supply.

By contrast, the Proposal for Digital Content Directive is to be welcomed. Digital content has sat uneasily in a divide between goods and services liability, but merits a *sui generis* regime that takes account of its particular characteristics. Most Member States do not have rules on this and even the UK Consumer Rights Act does not cover all the issues in the proposal. The UK Act maintains the repair and replacement terminology, whereas the Proposal simply talks about bringing the digital content into conformity and by its structuring of the remedy prevents a disproportionate cost remedy being imposed. In the online environment the difference between a repair and replacement may be less noticeable than for tangible goods, justifying the Proposal leaving the choice of how to bring the goods into conformity with the supplier. For tangible goods there is more justification for allowing the consumer to choose between repair and replacement, but this should be subject to controls where the costs are disproportionate. Ironically, the ways in which the Proposal for Digital Content Directive improves on the rules in the UK Act underlines the need for there to be flexibility for different approaches to be developed. This is particularly important in relation to new areas of regulation. If the first law had not been a UK Act, but rather a maximal harmonised directive, it would have prevented further regulatory innovation. This brings us back to the underlying philosophy of regulation. In this regard, the EU's role should be to set minimum standards and promote convergence, rather than impose solutions that may be logically coherent but have tensions with other equally defendable approaches to consumer protection.

THE PROPOSAL OF THE EU COMMISSION FOR A REGULATION ON ENSURING THE CROSS-BORDER PORTABILITY OF ONLINE CONTENT SERVICES IN THE INTERNAL MARKET*

Karl-Nikolaus Peifer

1. Introduction 163
2. Portability – Why Do We Have to Regulate It? 164
3. Why and How Do We Regulate Portability? 165
4. Supporters and Critics of the Draft Proposal. 166
5. The Core and Content of the Draft Proposal 167
   - 5.1. Overview 167
   - 5.2. Temporary Access (Articles 2 and 3) – How to Ensure it Works and Control its Use. 167
   - 5.3. Legal Fiction of Place of Use (Article 4) 168
   - 5.4. Influence on Contracts (Article 5) 169
   - 5.5. Services Covered. 169
   - 5.6. The Control Question 170
6. The Function of the Regulation within the Digital Agenda 171
7. Possible Effects 171

1. INTRODUCTION

Online content is often displayed on portable devices. It is technically accessible by any computer device with an access to electronic services. End users having a subscription have an interest to access services wherever users are and whenever they want to have access. Within the European Union this interest is protected by the freedom to move. The European Commission wishes to secure this

---

* The text was presented on 18 May 2016 at the Ferrara European Contract Law Seminar 'New Features of European Contract Law – Towards a Digital Single Market'. An expanded Italian version to be published in *AIDA* (*Annali Italiani del Diritto d'Autore*) is in preparation.

interest by imposing on service providers an obligation to grant access not only within the territory of the consumer's ordinary residence but also in territories of temporary residence. This obligation is the kernel of the proposal of a Regulation to ensure cross-border portability. Consumer organisations support the proposal, arguing that clients who have subscribed to services in their home country have a legitimate interest in also using the services when on vacation or when studying abroad. Service providers support the suggestion because they have an interest in levelling up the quality of their platforms. On the other hand, content rights holders put up resistance. They fear that the difference between temporary and ordinary residence remains unclear. Scenarios where this right could be misused are suggested. The overall fear is that the territoriality principle in copyright law and territorial licence schemes will be undermined. The text will discuss the future of the digital market within this seemingly small scope and its impact on other areas. It will focus on contract law and intellectual property rights. It argues that the proposal is a limited but necessary step forward from a legal policy point of view, which will help to build a unified digital market. In the long run, it will not only help to strengthen consumer trust but also foster new business models which are more tailored to a connected world.

2 PORTABILITY – WHY DO WE HAVE TO REGULATE IT?

Portability is about old facts in new clothes. Freedom of movement allows you to take your belongings with you wherever you go. In the digital world these belongings sometimes consist of having access to information which is made accessible on computer servers.[1]

Subscriptions to video download services in the present world very often grant ubiquitous and direct access only within the limits of contractual rights.[2] Contractual rights depend on licences from content providers. Content providers very often grant licences on a territorial basis, thus splitting up the unified digital market into licence territories. Multi-territorial licences are possible, but they are not readily granted. An impact assessment by the European Union found out that only 4% of online services are accessible Union-wide. This is not because service providers refrain from making their services more attractive to their customers, but because content providers refrain from licensing their content on a multi-territorial basis. Relying on the freedom of contract is therefore a slow route to take in moving forward to a digital unified market. This situation differs from a world where content is linked to tangible media, like CDs, DVDs or discs. While tangible media profit from the freedom of goods within the Union, digital

---

[1] See J. RIFKIN, *The Age of Access*, Putnam, New York 2000.
[2] Commission Staff Working Document: Impact Assessment, SWD(2015) 270 final, p. 8.

content rules are not fully governed by property and exhaustion rules.[3] One might leave this question open until content providers are willing to grant access and offer new business models. However, this position leaves consumers without alternative recourse and it will ossify the power to define for content providers how far the digital market can go.

Contract rules in the digital field define what content is, and how and for what purposes subscribers can access it. If contract only grants access in the territory where the licence was acquired, users of, for instance, a streaming service may only have access while in their home territory. Even if they have access abroad, the same content might not be provided. Freedom of contract allows these practices. Unfair competition law supports the position as long as the consumer is informed about the restrictions of his rights.[4] Even the fact that most contracts will be standard contracts does not help to label them as unfair terms if property rules (copyright rules) do not define what a consumer licence has to contain and where the misuse of the copyright property starts.[5] The Commission found that regulation is needed to protect the user's interest in getting access to services which the user has acquired with the expectation of being able to use it even while temporarily abroad, whether for a vacation or for other travel purposes. One might speak of a right to use Netflix while on a holiday. Technically, the Regulation partially defines intangible consumer rights to access content, but this right takes the form of cogent contract rules or contractual consumer rights. These contract rules aim to protect principles like 'What You Get Is What You See', therefore they define the legitimate interest of a consumer to effectively own what he has acquired.

3. WHY AND HOW DO WE REGULATE PORTABILITY?

One might wait for service providers to grant access in conformity with the terms of the contract. However, this depends on content owners being willing to partially give up territorial licensing or at least to grant Union-wide licences at affordable prices. The Commission clearly stated that it does not have too much hope of seeing this happen.[6] Regulation therefore is used to push or 'nudge' services providers to grant more access. As the Union has limited power to re-define national copyright rules,[7] it might take the big step of establishing

---

[3] See Case 78/70, *Polydor* [1971] ECR 487; Case 270/80, *Polydor v. Harlekin* [1982] ECR 329.
[4] See Arts. 6 and 7 of the Directive 2005/29/EC (Unfair Commercial Practices Directive) on misleading practices including misleading omissions.
[5] Art. 3(1) Directive 93/13/EC (Unfair Terms Directive).
[6] Commission Draft Proposal, COM(2015) 627 final, p. 5 (Draft Proposal).
[7] See Art. 345 TFEU.

a Community Copyright Title;[8] however, this will not necessarily abolish national copyright rules and with them territorial licensing based on national rules. This explains why the Union choses consumer contract law to protect the consumer's interest in owning and carrying with him what he wanted to acquire.

Regulation of portability is at Union level and is achieved proposing proposed a regulation not a directive. Why is that so? A leading example is the *Karen Murphy* copyright case.[9] The British pub owner Murphy acquired access to satellite signals of sports events (the British Premier League) by obtaining a decoder device that had been lawfully licensed in Greece, but only for Greece. The ECJ ruled that a device that had lawfully been distributed in Greece may move freely within the Union. Intellectual property rights incorporated in the concrete device are exhausted once this device has been legally sold. The same is true for used software.[10] The ECJ cases in this area show that there is an interest in granting Union-wide access to digital content. The ECJ decisions, however, regulate specific cases only; a regulation would regulate the phenomenon as such with Union-wide and immediate effect.

4. SUPPORTERS AND CRITICS OF THE DRAFT PROPOSAL

Industry, especially the content industry, is not happy with the proposal. They would have preferred to use self-regulatory solutions.[11] User associations support the Commission's idea simply because they doubt that voluntary licensing of Union-wide access will happen quickly and uniformly.[12] The Commission strengthened the position of the consumer The draft Portability Regulation is therefore a consumer protection device. It limits freedom of contract and has some impact on intangible property rules in the copyright and neighbouring rights sector.[13]

The draft Regulation covers only audiovisual content (Art. 2(e)), thus video streaming services such as Netflix and Amazon, as well as YouTube. It may have side effects for music streaming services such as Spotify, for which the Union has introduced a special instrument to support multi-territorial licences.[14] Further side effects relate to free-to-air television services, which will be covered if they

---

[8] Art. 118 TFEU.
[9] Case C-403/08 and 429/08, *Karen Murphy*, 4 October 2011.
[10] Case C-128/11, *Used Soft*, 7 July 2012.
[11] See Draft Proposal, *supra* n. 6, p. 5.
[12] Draft Proposal, *supra* n. 6, p. 5.
[13] See recitals 4–6 Draft Proposal, *supra* n. 6.
[14] See Title III of the Directive 2014/26/EU of 24 February 2014 on collective management of copyright and related rights and multi-territorial licensing of rights in musical works for online uses in the internal market.

require users to register (Art. 2(e)(2)).[15] The draft Regulation directly aims to protect consumers who are travelling and away from home for a short period in time. However, it is feared that the Regulation is will have spill-over effects and the wording supports this fear because the Regulation does not define what temporary access is and how it is limited. The possibility cannot be excluded that social media such as Facebook and search engine providers such as Google are covered by the Regulation.

5. THE CORE AND CONTENT OF THE DRAFT PROPOSAL

5.1. OVERVIEW

The text of the Regulation is relatively short. It has only seven articles, but many more recitals (29). Article 3 is the core provision. Article 4 describes the technique to grant overall and horizontal effect and clear out copyright problems. Article 4 is the instrument which has the closest link to contract law. Article 2 restricts the reach of the Regulation to temporary use outside the licence territory.

5.2. TEMPORARY ACCESS (ARTICLES 2 AND 3) – HOW TO ENSURE IT WORKS AND CONTROL ITS USE

The Regulation provides for a contractual obligation to grant temporary access to consumers in territories outside their licence territory (Art. 3(1)). The Regulation does not directly define what temporary access is. By way of a negative definition it states that 'temporary' means everything which happens outside the Member State of residence. Therefore, temporary means 'not regular'. Everything that is outside the resident state is therefore temporary. This exceeds some of the arguments used to introduce the contractual instrument. The Commission originally defended its plans by citing the wish to grant access rights for travellers. The text of the Regulation, however, seems to extend it to those abroad for living part of the year (for example, northern Europeans spending the winter months in the Mediterranean, southerners moving to the cooler areas in summer), or students or academics spending months or even a year abroad and taking their services with them.

This solution will not be problematic for contracts granting paid access. This is what the Regulation is mainly concerned with. Once the user has paid,

---

[15] Draft Proposal, *supra* n. 6, p. 4.

the content provider should not worry about whether the user watches films or sports events at home or abroad. Even if the user pays with his or her attention to advertising, the advertiser will prefer being able to also reach the consumer abroad. A different problem arises if the user has more than one factual residence and therefore has the opportunity to 'steal' a cheaper contract by purchasing the service in the cheaper-priced territory. The fear that this will happen seems, however, to a large extent exaggerated because barriers to moving are both high and costly so it does not seem very realistic that people move just for the purpose of getting cheap access to films or sports events.

A different point is the problem of grey markets for access tools. The *Karen Murphy* case has shown that devices to access British content could be bought at cheap prices in Greece and then still be used in British pubs. The Commission will have to pay close attention to misuse scenarios and may have to build in defence mechanisms against such misuse. Moreover, contract providers will be given a right to force service providers to ensure that only residents are granted access abroad (see section 6.)

5.3. LEGAL FICTION OF PLACE OF USE (ARTICLE 4)

Article 4 establishes the connection between copyright and contract law. A contract may allow temporary access outside the licence territory. Copyright law still holds that such a use may be a copyright violation. To avoid this interpretation, Article 4 establishes a legal fiction that a temporary use of content outside the licence territory is deemed to take place within the licence territory and that therefore there is no copyright violation. This legal fiction seems to be something like a 'sleight of hand'; however, the fiction is not new. It was used in the Cable and Satellite Directive to define the place where content is used when being 'broadcast' by satellite. The Cable and Satellite Directive states that the broadcast is executed only in the territory of the uplink territory. This means that contractual rights to uplink signals have to be acquired in the uplink territory only. The Portability Regulation uses this logic to define the place where contractual rights are executed. They are executed at the place of registration only.

The main argument against this definition is the territoriality principle in copyright law. As this principle conflicts with the freedom of movement (and the freedom of services), property and liberty principles collide. This is a common conflict when property rules and antitrust rules collide. In that field, the collision is solved by defining whether a certain use belongs to the essence of the property right. Licence and contractual rights, however, are not harmonised at Community level. This supports the argument that the Regulation does not re-define the essence of a property right if the instrument is restricted to contract holders only.

5.4. INFLUENCE ON CONTRACTS (ARTICLE 5)

While Article 3 defines contractual rights and Article 4 clarifies where a contractual use is located, an additional provision is needed to combine the contractual relationships between users and internet service providers on the one side and service providers and content providers (rights holders) on the other. This is achieved by Article 5. Article 5 states that contractual provisions between rights holders and service providers which hinder the access rights of consumers are unenforceable. Content providers may use such provisions, but the service provider cannot enforce them *vis-à-vis* the consumer. Article 7 holds that this mechanism also applies to existing contracts.

The mechanism has a flash of genius; however, it does not help if rights holders refrain from licensing their content to service providers in the first place. Here is a loose thread in the system. The regulation relies on the expectation that the big majors outside the Community will not refrain from licensing content to service providers in the Union. This expectation is based on the hope that the Community is too attractive to be ignored. The expectation might, however, also induce right holders to licence their content only on a Community-wide basis. It is not unrealistic to expect that those licences will be much more costly than territorial licences. Smaller service providers might feel reluctant to license content on that (more costly) basis at all. Those market reactions therefore might work against the current Community plans.

5.5. SERVICES COVERED

The Regulation covers 'online content services'. It is directed at paid subscriptions but includes subscriptions by which the user does not pay with money. Article 2(e) defines 'online content service' as the lawful provision of services 'to a subscriber on agreed terms either (1) against payment of money; or (2) without payment of money'. The latter case covers free services such as YouTube or other social media providing audiovisual content. However, these services are only covered if 'the subscriber's member state of residence is verified by the provider'. Therefore, services such as YouTube are partly covered, while free-to-air television services are only covered if they control access by their users.

This raises two problems: if rights holders expect free-to-air audiovisual service providers to be able control the where users access their services (Article 5) then these service providers need access control tools. If they use these tools, they are covered by the Regulation. This produces the next problem of those free-to-air providers who have a legal obligation to keep access unrestricted, such as is the case for German public television. Public television broadcasters in Germany have to take safeguards for granting unrestricted access to users in Germany which means that their video streaming services ('*Mediatheken*') do not contain any tools to control whether access is requested by legitimate users. As any household in Germany pays a contribution (so-called *Rundfunkbeitrag*) to finance the public broadcasting system as such, there is no need to control access. Data protection laws would not regard access controls as 'necessary' to derogate from the consent rule.

5.6. THE CONTROL QUESTION

A major problem is caused by the control mechanisms which the Regulation allows with regard to the place of residence of users privileged by a virtually extended licence. Article 5(1) states that restrictions to digital access by contracts are unenforceable. This means a horizontal interference with all contracts including those that have been concluded in the past and between parties outside the territorial reach of the Regulation. To soften this intervention, rights holders 'may require that the service provider make use of effective means in order to verify that the … service is provided … in conformity with Article 3(1)'. This is meant to protect rights holders against misuse. A misuse would be where a user is granted access outside the territory of his residence provided this user is not only temporarily abroad. Therefore, a service provider may be imposed with the obligation to check their users' place of habitual residence.

The Regulation does not set out any criteria for how this may be done; it simply states that control mechanisms must be 'reasonable' and 'necessary' to achieve their purpose. Recital 23 states that 'technical and organizational measures' including 'sampling of IP addresses' may be imposed; however, 'precise location data' are expressly excluded by the recitals. Therefore, the Commission proposal leaves the question quite open and very much subject to the freedom of contract. A Council document of 31 March 2016[16] has added a list of up to eight default criteria to identify users that may be used in the absence of an industry standard or practice, among them the identity card, a billing address, bank details, the place of installation of a set top box, the internet or telephone contract details granting access and the licence contract. This list looks like a wish list from the standpoint of a rights holder and like a disaster from the perspective of a data protection agency. It is highly problematic to leave the list to the industry. The proposal still restricts excessive use of personal data by implying that any use of such data has to be necessary and serve the purpose of checking the user's habitual residence. Therefore it may still be in line with the General Data Protection Regulation (Arts. 5(1)(b)

---

[16] <http://g8fip1kplyr33r3krz5b97d1.wpengine.netdna-cdn.com/wp-content/uploads/2016/04/Online-content-portability.pdf>.

and 6(1)(c) GDPR).[17] However, it remains unclear whether only some or all of these criteria might be chosen by a service provider. The necessity requirement and the principle of data minimisation (Art. 5(1)(c) GDPR) both call for more modesty than the Council document seemingly wants to allow. Therefore, the Regulation should more rigidly restrict the criteria to be used. Otherwise data protection agencies might force to do what the Regulation should have clarified in the first place.

6. THE FUNCTION OF THE REGULATION WITHIN THE DIGITAL AGENDA

The number of recitals and the effort spent on the impact assessment indicates that the project has a central role within the Digital Agenda. There have been several efforts to attack the territoriality principle,[18] with some voices calling for a technologically neutral approach to the public communication of television programmes via streaming.[19] The Portability Regulation is therefore a kind of trial balloon with a still limited impact on business models in copyright law. It is also, however, a cornerstone of the plans regarding digital contracts. The plans help to define contractual rights in relation digital content more clearly. Within the realm of digital content, property rights are usually defined by the rights holders through contracts. Consumers very often have to take what is given to them. Mandatory access rights are rare. The draft Regulation is therefore also an instrument to define intellectual property rights within licensing schemes, one of the few areas in which no harmonisation has taken place so far.

7. POSSIBLE EFFECTS

If the Regulation comes into force, contractual rights of consumers will be much more clearly defined. Service providers will be able to offer more attractive services to their clients. There is some hope within the Commission

---

[17] The General Data Protection Regulation (GDPR) [2016] OJ L119/1 was released on 16 April 2016 and will be applicable from 25 May 2018; see Article 99(2) GDPR.

[18] See the so-called 'Reda Report': Draft Report on the implementation of Directive 2001/29/EC of the European Parliament and of the Council of 22 May 2001 on the harmonization of certain aspects of copyright and related rights in the information society (2014/2256(INI)), 11 June 2015.

[19] See the EU consultation on the Cable and Satellite Directive <https://ec.europa.eu/digital-agenda/en/news/consultation-review-eu-satellite-and-cable-directive#English>. German public broadcasters strongly support the ideal of a technologically neutral communication right for free-to-air-TV, see German Bundesrat, Decision of 3 May 2013, No. 1 of the Bundesrats-Drucksache no. 265/13; C. KROGMANN, *Zeitschrift für Urheber- und Medienrecht*, Nomos, Baden-Baden 2013, p. 457.

that these two effects will 'boost the economy' by creating more diverse digital markets. However, some of the hope relies on rights holders being disposed to offer the content needed to feed services. There is some fear that major movie companies are neither prepared nor willing to grant territorial licences which are then expanded to cross-border uses. There is no tool to force rights holders to do so. Service providers therefore might run out of licences. The Commission hopes that the European market is too attractive to stop feeding content into it. However, the Commission will have to watch closely how and to what extent market practices change once the Regulation enters into force. This again shows that the draft Regulation is a 'trial balloon'. Technically, it is a clever attempt to push forward the idea of a single digital market because it is restricted to relatively marginal cases, but might nevertheless have a remarkable effect for consumers.

THE LAW APPLICABLE TO CONSUMER CONTRACTS IN THE DIGITAL SINGLE MARKET

Peter KINDLER

1. The Substantive Law Background 173
2. The Law Applicable to Consumer Contracts: General Outline 175
3. The Key Connecting Factor: Activities 'Directed' to the Consumer Country (Article 6(1)(b) Rome I Regulation)...................... 176
    3.1. The Relevant Facts under Article 6(1)(b) Rome I Regulation: General Outline 176
    3.2. Distribution Activities Carried out by an Independent Third Party.. 177
    3.3. Advertising of Products in the Consumer Country 178
    3.4. The Role of the Professional's Website 179
4. Merely Indicative Facts 182
    4.1. Conclusion of the Contract at a Distance..................... 182
    4.2. Causative Connection 182
5. The Need for Specific Conflicts of Law Rules for International Consumer Contracts in the Digital Single Market.................. 183

1. THE SUBSTANTIVE LAW BACKGROUND

In its 2015 proposal for a Directive on certain aspects concerning contracts for the supply of digital content,[1] the Commission stated that 'nothing in this Directive should prejudice the application of the rules of private international

---

[1] COM(2015) 634 final; on this proposal see M. SCHMIDT-KESSEL, K. ERLER, A. GRIMM and M. KRAMME, 'Die Richtlinienvorschläge der Kommission zu Digitalen Inhalten und Online-Handel' (2016) *Zeitschrift für das Privatrecht der Europäischen Union (GPR)* 1; M. WENDLAND, 'GEK 2.0? Ein europäischer Rechtsrahmen für den Digitalen Binnenmarkt – Die neue Richtlinie über Verträge zur Bereitstellung digitaler Inhalte (Digitalgüter-Richtlinie)' (2016) *GPR* 8 et seq.; L. PRATS ALBENTOSA, 'Comercio electrónico europeo de bienes y contenidos digitales: mayor armonización de los derechos de los consumidores y de los deberes de los oferentes' (16.02.2016) *Diario La Ley* No. 8703, 1 et seq.

law, in particular Regulation (EC) No 593/2008 (Rome I) and Regulation (EC) No 1215/2012 (Brussels I a).'[2] The same is true for the 2015 Proposal for a Directive on certain aspects concerning contracts for the online and other distance sales of goods.[3] In the eyes of the Commission, the relevant private international law instruments were adopted quite recently and the implications of the Internet were considered closely in the legislative process. The Commission thinks that the existing conflict-of law rules take specific account of Internet transactions, in particular those on consumer contracts. In fact, the rules laid down in Rome I (Article 6) aim at protecting consumers, *inter alia*, in the Digital Single Market by giving them the benefit of the non-derogable rules of the Member State in which they are habitually resident. But they do not specifically address the phenomenon of online advertising and/or the online formation of contracts. Nevertheless, according to the Commission, together with the proposed contract rules for the purchase of digital content,[4] the existing rules on private international law establish a clear legal framework for buying and selling in a European digital market, which takes into account both consumers' and businesses' interests. Therefore, the legislative proposals aiming at the creation of a digital single market do not require any changes to the current framework of EU private international law,[5] including the Regulation (EC) 593/2008 (Rome I).[6]

At a substantive law level, with its proposals for two directives on the supply of digital content and online sales of goods, the Commission provides a comprehensive legal framework for a private law in the digital world. The rules for the supply of digital content enter to a large extent legal *terra incognita* and are therefore ground-breaking. The rules for online sales of goods result in the full harmonisation of large parts of the consumer warranty law and therefore have an extensive impact on the sales law of the Member States. The proposals are characterised by a relatively high level of consumer protection – with the exception of the high requirements for rescission in the case of lack of conformity of the digital content with the contract.[7]

---

[2] Recital 49.
[3] COM(2015) 635 final, recitals 6 and 7, 37. On this proposal see WENDLAND, *supra* n. 1, p. 8 et seq.; M. WENDLAND, 'Ein neues europäisches Vertragsrecht für den Online-Handel? Die Richtlinienvorschläge der Kommission zu vertragsrechtlichen Aspekten der Bereitstellung digitaler Inhalte und des Online-Warenhandels' (2016) *EuZW* 126 et seq.
[4] See *supra* n. 1.
[5] COM(2015) 634 final, pp. 3–4.
[6] In particular, the Commission refers to the detailed explanation of the EU rule on applicable law and jurisdiction in the Digital Single Market in Annex 7 to the 'Commission Staff Working Document containing the Impact Assessment accompanying the Proposals for a Directive of the European Parliament and of the Council on certain aspects concerning contracts for the supply of digital content and a Directive of the European Parliament and of the Council on certain aspects concerning contracts for the online and other distance sales of goods', SWD(2015) 275.
[7] In detail, see WENDLAND, *supra* n. 3, p. 126 et seq.

2. THE LAW APPLICABLE TO CONSUMER CONTRACTS: GENERAL OUTLINE

In view of this proposed legislative self-restraint as to private international law matters, it seems all the same useful to analyse the existing private international law rules and the relevant case law of the CJEU, in order to assess the usefulness and meaningfulness of the Commission's approach.

Currently, pursuant to Article 6 Rome I Regulation, a consumer contract concluded by a natural person for a purpose which can be regarded as being outside his trade or profession (the consumer) with another person acting in the exercise of his trade or profession (the professional) shall be governed by the law of the country where the consumer has his habitual residence, provided that the professional: (a) pursues his commercial or professional activities in the country where the consumer has his habitual residence, or (b) by any means, directs such activities to that country or to several countries including that country, and the contract falls within the scope of such activities.[8] This so-called targeted activity test[9] follows the example of Article 17(1)(c) Brussels I bis Regulation/ Article 15(1)(c) Brussels I Regulation. It is a more general and technology-neutral reformulation of the preconditions already contained in Article 5(2) Rome Convention and Article 13(3) Brussels Convention respectively.[10] Except for the less restrictive wording, meant to make the operation of the provision easier in electronic commerce, no substantial change was made.[11] The legislator had no intention to narrow the scope of the provision to the detriment of consumers: this is why Article 6 Rome I neither requires contracts to be concluded at a distance,[12] nor burdens the consumer with the proof of a causative connection between the professional's 'targeted' activity and the conclusion of the contract.[13]

---

[8] Case C-190/11, *Mühlleitner v. A. Yusufi/W. Yusufi*, para. 36; Case C-218/12, *L. Emrek v. V. Sabranovic*, para. 22. Pursuant to Article 6(2) of the Regulation, the parties may choose the law applicable to a contract which fulfils the requirements of paragraph 1, in accordance with Article 3. Such a choice may not, however, have the result of depriving the consumer of the protection afforded to him by provisions that cannot be derogated from by agreement by virtue of the law which, in the absence of choice, would have been applicable on the basis of paragraph 1.

[9] In detail, see G. CALLIESS, *Rome Regulations*, Wolters Kluwer, Alphen aan den Rijn 2015, Rome I Article 6, Consumer Contracts, marginal no. 41.

[10] On this cf. P.A. NIELSEN, 'Art. 15 Brussels I Regulation' in U. MAGNUS and P. MANKOWSKI (eds.), *Brussels I Regulation*, 2nd ed., Sellier, Munich 2012, para. 30; J. HILL, *Cross-Border Consumer Contracts*, Oxford University Press, Oxford 2009, pp. 4–38. Cross-border electronic commerce contracts were already covered by Art. 5(2) first indent Rome Convention: P. MANKOWSKI, 'Das Internet im Internationalen Vertrags- und Deliktsrecht' (1999) *RabelsZ* 63, 203 et seq.; cf. A.V. DICEY, J. MORRIS and L. COLLINS, *The Conflict of Laws*, 15th ed., Sweet and Maxwell, London 2012, paras. 33-133 et seq.; S. LEIBLE, 'Binnenmarkt, elektronischer Geschäftsverkehr und Verbraucherschutz' [2010] *JZ* 272.

[11] CALLIESS, *supra* n. 9, Rome I Article 6, Consumer Contracts, marginal no. 15 et seq.

[12] Case C-190/11, *Mühlleitner v. A. Yusufi/W. Yusufi*, paras. 37 et seq., 40, 42.

[13] Case C-218/12, *L. Emrek v. V. Sabranovic*, paras. 24 et seq.

The practical relevance of Article 6 Rome I becomes clear when we take into account that national provisions transposing the Union legislation on consumer contract law still significantly diverge today on essential elements of a sales contract, such as the absence or existence of a hierarchy of remedies, the period of the legal guarantee, the period of the reversal of the burden of proof, or the notification of the defect to the seller.[14] Furthermore, even for the contracts covered by the proposed directives there is no full harmonisation (despite Art. 3 draft Directive on online sales and Art. 4 draft Directive on digital goods):[15] divergent national rules may still exist on limitation periods (cf. Art. 14 draft Directive on online sales), on commercial guarantees (cf. Art. 15(4) draft Directive on online sales), on the consumer's right to damages (cf. Art. 14(2) draft Directive on digital goods), on the seller's right of redress (cf. Art. 16 draft Directive on online sales; Art. 17 draft Directive on digital goods).

3. THE KEY CONNECTING FACTOR: ACTIVITIES 'DIRECTED' TO THE CONSUMER COUNTRY (ARTICLE 6(1)(B) ROME I REGULATION)

3.1. THE RELEVANT FACTS UNDER ARTICLE 6(1)(B) ROME I REGULATION: GENERAL OUTLINE

The generally consumer-friendly tendency of Article 6 Rome I Regulation is also the guiding theme of the relevant case law of the Court of Justice. The decisive question in these cases is always: under which circumstances can one say that the professional is 'directing' his activities to the consumer's country? This is the main factual element, the key connecting factor, which leads to the application of the substantive law of the country where the consumer has his habitual residence. Generally speaking, Article 6(1)(b) applies where the professional (including his staff and affiliates) is not physically present in the consumer's country to solicit the contract, but where he *directed his commercial or professional activities* to that country by any means in order to solicit contracts with consumers from the country. Thus, the crucial factual element is the professional's *intention* to direct activities to a certain country: it justifies, *vis-à-vis* the professional, the

---

[14] Draft Directive on online sales, recitals 5–8.
[15] Draft Directive on the online sale of goods, p. 7 of the Explanatory Memorandum: 'The proposal will not harmonise all aspects concerning contracts for the online and other distance sales of goods'; Article 3(9) draft Directive on digital contents: 'In so far as not regulated in this Directive, this Directive shall not affect national general contract laws such as rules on formation, the validity or effects of contracts, including the consequences of the termination of a contract.'

application of the law of this very country.¹⁶ Such a marketing strategy of the professional can only be assessed from objective criteria which allow conclusions on the professional's intention to be drawn.

In this regard, the fact that a contract was concluded with a consumer from that country is one factual indicator among others for such intention.¹⁷ In other words, the contract itself does not constitute the required 'directed activity' but rather is a result of such marketing activity. Obviously, the mere fact that a consumer contract was concluded does not as such fulfil the prerequisites of this connecting factor: otherwise the 'directing' requirement of Article 6(1)(b) would be redundant.¹⁸ The provision covers soliciting of the conclusion of contracts by any means, including a website. The irrelevance of the conclusion of the contract as such follows, *inter alia*, from the further requirement of Article 6(1) that 'the contract falls within the scope of such activities.' All in all, the court has to conclude from all the circumstances of the individual case whether the professional has 'targeted' the consumer country or not.¹⁹ The precise meaning of this formula can be understood only in light of the relevant case law.

3.2. DISTRIBUTION ACTIVITIES CARRIED OUT BY AN INDEPENDENT THIRD PARTY

The Court of Justice of the European Union has had several occasions to establish under which specific circumstances, in the light of Article 15(1)(c) Brussels I Regulation,²⁰ one can say that a professional 'directs' his activities to the country where the consumer has his habitual residence. Firstly, there can be hardly any doubt that a professional directs commercial or professional activities to the consumer's country where he contracts the distribution activities out to an independent third party which itself by virtue of physical presence pursues activities in the consumer's country within the scope of Article 6(1)(a) Rome I Regulation. The relevant leading European case is *Pammer*, where Peter Pammer, a consumer residing in Austria, had booked a voyage on a freighter of a German shipping company (Reederei Karl Schlüter).²¹ Mr Pammer booked the voyage through a German intermediary company (Pfeiffer) which offered freighter

---

16   Higher Regional Court of Munich, case 15 U 2341/15 Rae, (2016) *IPRax*, vol. 3 (May/June), p. X, also in (2016) *BeckRS* 06263.
17   Case C-190/11, *Mühlleitner v. A. Yusufi/W. Yusufi*, para. 44; Case C-218/12, *L. Emrek v. V. Sabranovic*, para. 28.
18   See Case C-585/08 and C-144/09, *Alpenhof v. O. Heller / P. Pammer v. K. Schlüter*, para. 76; J. ØREN (2003) 52 *ICLQ* 665; L. GILLIES, 'Addressing the "Cyberspace Fallacy": Targeting the Jurisdiction of an Electronic Consumer Contract' (2008) *Int. J. L. & Info. Tech.* 242, 254 et seq.
19   Case C-585/08 and C-144/09, *Alpenhof v. O. Heller / P. Pammer v. K. Schlüter*, para. 92.
20   This provision served as a model for Article 6 Rome I Regulation, see *supra* n. 2.
21   Case C-144/09 *P. Pammer v. K. Schlüter*.

journeys on its website on the Austrian market. The intermediary company described the voyage on its website, indicating, *inter alia*, that there was a fitness room, an outdoor swimming pool, a saloon and video and television access on the vessel. Mr Pammer refused to embark and sought reimbursement of the sum which he had paid for the voyage, on the grounds that that description did not, in his view, correspond to the conditions on the vessel. Since Reederei Karl Schlüter reimbursed only part of that sum, Mr Pammer claimed payment of the balance, before an Austrian court. Reederei Karl Schlüter contended that it did not pursue any professional or commercial activity in Austria and raised the plea that the court lacked jurisdiction. The CJEU granted jurisdiction of the Austrian courts under Article 15(1)(c) Brussels I Regulation because the voyage contract was a consumer contract, and held that the intermediary company had engaged in advertising activity in Austria *on behalf of* Reederei Karl Schlüter by means of the Internet. The Court held that the fact that the website was the intermediary company's and not the trader's site does not preclude the trader from being regarded as directing its activity to other Member States, including that of the consumer's domicile, since that company was acting for and on behalf of the trader.[22] Unfortunately, the case file did not reveal sufficient facts on the mode and intensity of the cooperation between the intermediary and the shipping company.

3.3. ADVERTISING OF PRODUCTS IN THE CONSUMER COUNTRY

Secondly, there are cases where the conclusion of the contract was preceded by the professional advertising his products in the consumer country or in several countries including that country. Such advertising can be conducted by the traditional offline means of sending out letters and catalogues, buying advertisements in print media or booking spots on the radio or television. In order to determine the countries thus targeted, it is helpful to refer to the Giuliano-Lagarde Report which reads: 'If, for example, a German makes a contract in response to an advertisement published by a French company in a German publication, the contract is covered … If, on the other hand, the German replies to an advertisement in American publications, even if they are sold in Germany, the rule does not apply unless the advertisement appeared in special editions of the publication intended for European countries. In the latter case the seller will have made a special advertisement intended for the country of the purchaser.'[23] The general rule underlying this example can be summarised as follows: the court should rely on the media data of the advertising channel used

---

[22] *Ibid.*
[23] M. GIULIANO and P. LAGARDE, *Report on the Rome Convention* [1980] OJ C282/1, 24.

with regard to its diffusion rate, but accidental spill-over effects do not constitute targeted activity.[24]

The same rule applies, in principle, for advertising in online media, where of course the potential spill-over effects might be more significant. Thus, where the professional sends out (spam) e-mails, or otherwise engages in advertising online, these activities are targeted to *all* countries where its receivers can reasonably be expected to be habitually resident. It is up to the professional to prove that he took due diligence in limiting his activities and that a certain consumer was addressed only accidentally by virtue of spill-over effects.[25]

3.4. THE ROLE OF THE PROFESSIONAL'S WEBSITE

Finally, several cases concern the question what role a website of the professional plays in the targeted activity test.[26] In this regard, the joint declaration quoted in recital 24 Rome I Regulation states that 'the mere fact that an Internet site is accessible is not sufficient ... although a factor will be that this Internet site solicits the conclusion of distance contracts and that a contract has actually been concluded at a distance, by whatever means'. The test of whether a website solicits the conclusion of distance contracts in the consumer country has to be made from the objective perspective of an ordinary Internet user taking into account all circumstances of the individual case.[27] But what does this formula mean in the single case? In *Pammer* the CJEU made clear that such circumstances include – but are not limited to – the content of the webpage, for example if the telephone number includes a country dialling code, if travel directions are given for consumers from abroad, if different languages can be selected, which top-level domain is used, etc.[28]

In case of a trading website, i.e. an interactive website, which allows consumers to transfer an offer to the professional,[29] the professional's intention to 'direct' its activity toward the consumer country is obvious, for example, if the consumer is asked for his address, if special options for overseas payments and delivery are available, if the website includes information on customs or

---

[24] CALLIESS, *supra* n. 9, Rome I Article 6, Consumer Contracts, marginal no. 49.
[25] Opinion of Attorney General Trstenjak in joint cases C-585/08 and C-144/09, *Alpenhof v. O. Heller / P. Pammer v. K. Schlüter*, para. 88 (on the risks of sending out spam); P. MANKOWSKI, 'Die Darlegungs- und Beweislast für die Tatbestände des Internationalen Verbraucherprozess- und Verbrauchervertragsrechts' (2009) *IPRax* 474 et seq.
[26] This aspect was highly controversial in the legislative process: see CALLIESS, *supra* n. 9, Rome I Article 6, Consumer Contracts, marginal no. 14 et seq.
[27] Opinion of Attorney General Trstenjak in Joint Cases C-585/08 and C-144/09, *Alpenhof v. O. Heller / P. Pammer v. K. Schlüter*, para. 76; S. LEIBLE and M. MÜLLER, 'Die Bedeutung von Websites für die internationale Zuständigkeit in Verbrauchersachen' (2011) *NJW* 495, 496 et seq.
[28] Case C-144/09, *P. Pammer v. K. Schlüter*, paras. 81 et seq.
[29] On this type of website see recitals 38 and 39 Directive 2011/83/EU on consumer rights.

value added tax specific to the consumer country, if the professional's terms and conditions include choice of law and jurisdiction clauses, etc. In addition, the track record of the professional is a relevant factor, i.e. the quantity and intensity of business relations with consumers from a certain country.[30] On the other hand, the mere fact that contact details and an e-mail address are given on a website is not evidence capable of demonstrating the professional's intention to direct his activity toward another country: this kind of information is compulsory under Article 5(1) of Directive 2000/31/EC on electronic commerce.[31]

Let's take the example of the US-based Internet warehouse Amazon. The company runs its corporate group activities in Europe through different-language websites such as amazon.co.uk, amazon.de, and amazon.it.[32] These domains are used as trading names for Luxembourg-based Amazon EU Sàrl, which is the sole contracting partner for all contracts concluded through these websites. There is not doubt that each of these sites is 'directed' to the consumers habitually resident in the country of the top-level domain that is used. However, as each of the sites also deliver consumers resident in other countries, the marketing strategy of Amazon is oriented along cultural spheres and language boundaries rather than political frontiers, which is a very good example of how e-commerce transcends the classical localisation criteria used in private international law.[33] This represents precisely the circumstances covered by the wording of Article 6(1)(b) Rome I Regulation, i.e. an activity that is directed to several countries including that country. The law of the consumer country applies if the professional follows an international marketing strategy directed to an international audience. It is not necessary for the website to target only or specifically the consumer country in a given case.

Another example concerns the so-called passive website:[34] in this case the website merely offers information on the trader, his goods and/or services and his contact details.[35] In general, the professional has no intention of directing his activities to foreign countries. Let us take the example of a consumer – Mr Atzmann – residing in Germany. Mr Atzmann adores English shortbread cookies. Surfing on the Internet, Mr Atzmann spots the catalogue of a trader residing in the UK.[36] Using e-mail, Atzmann orders a huge quantity of cookies, which are delivered by an international parcel service. When submitting his

---

[30] Joint Cases C-585/08 and C-144/09, *Alpenhof v. O. Heller / P. Pammer v. K. Schlüter*, paras. 93 et seq.; cf. R. PLENDER and M. WILDERSPIN, *European Private International Law*, Sweet and Maxwell, London 2014, pp. 9–56 et seq.

[31] [2000] OJ L178/1; see Joint Cases C-585/08 and C-144/09, *Alpenhof v. O. Heller / P. Pammer v. K. Schlüter*, para. 78.

[32] CALLIESS, *supra* n. 9, Rome I Article 6, Consumer Contracts, marginal no. 51.

[33] *Ibid.*, Rome I Article 6, Consumer Contracts, marginal no. 51.

[34] The example is based on B. v. HOFFMANN and K. THORN, *IPR*, Beck, Munich 2007, §10 marginal no. 73a.

[35] See the definition in recital 20 Directive 2011/83/EU on consumer rights.

[36] P. KINDLER, *Einführung in das neue IPR des Wirtschaftsverkehrs*, Verlag Recht und Wirtschaft, Frankfurt am Main 2009, pp. 48–49.

order, Atzmann accepted a choice-of-law clause according to which the sales contract was subject to the laws of the UK. Can Atzmann invoke the protection granted by German consumer laws (for example §305 et seq. BGB on standard business terms)? In a similar case, the Higher Regional Court of Karlsruhe refused to apply Article 6 Rome I Regulation because the trader did not 'direct' its activities to Germany.[37] This is in line with the joint declaration quoted in recital 24 Rome I Regulation according to which a 'passive website' does not prove the trader's intention to direct his activities to an international audience.[38] To be sure, the trader can, at any rate, explicitly exclude certain countries from his activities.[39]

The relevant facts were less clear in the CJEU Case C-144/09 *Alpenhof*, where a consumer residing in Germany searched the Internet for a hotel in Austria. He found the website of Hotel Alpenhof and made a request to the e-mail address given on the website. He received an offer which he accepted via e-mail. In this case, the accessibility of the website and the e-mail address strongly suggest a targeted activity. Furthermore, the fact that the hotel afterwards responded to the e-mail request of the German consumer (whose habitual residence was recognisable to the hotel) and concluded a contract at a distance with him creates a *prima facie* presumption that the hotel directed commercial activities to German consumers. In cases like this, it is the responsibility of the trader to argue it did not intend to direct commercial activities to the German market and to prove what kind of ring-fencing mechanisms – if any – were operative to exclude German customers if that market should not have been targeted.[40] One should note, however, that the hotel contract is excluded from the scope of Article 6 Rome I Regulation by virtue of paragraph 4(a) of that Article.

In short, the following list of relevant connecting facts has to be taken into consideration according to the CJEU in *Pammer*:[41]

- the international character of the services offered (e.g. a freighter journey);
- travel directions for consumers from abroad to the trader's central administration/place of business on the website;[42]

---

[37] OLG Karlsruhe 24.08.2007 (2008) *NJW* 85 = (2008) *IPRax* 348; S. LEIBLE and M. LEHMANN, 'Die Verordnung über das auf vertragliche Schuldverhältnisse anzuwendende Recht ("Rom I")' (2008) *RIW* 528, 538.
[38] Case C-190/11, *Mühlleitner v. A. Yusufi/W. Yusufi*, para. 19.
[39] P. MANKOWSKI, 'Neues zum "Ausrichten" unternehmerischer Tätigkeit unter Art. 15 I lit. c. EuGVVO' (2009) *IPRax* 238 ff.
[40] With regard to the burden to argue and prove see generally P. MANKOWSKI, *supra* n. 39, p. 483 et seq.; for further implications of the CJEU decision in Joint Cases C-585/08 and C-144/09, *Alpenhof v. O. Heller / P. Pammer v. K. Schlüter* see S. LEIBLE and M. MÜLLER, *supra* n. 27, p. 495.
[41] Case C-144/09, *P. Pammer v. K. Schlüter*.
[42] On the importance of travel directions under the targeted activity test see German Federal Court of Justice, 24.04.2013 (2013) *RIW* 563, marginal no. 22.

- the use of different languages or currencies than the ones usually adopted in the Member State where the trader resides together with the possibility, for the consumer, to book in that language;[43]
- the indication of telephone numbers including a country dialling code which foreign consumers have to dial;
- the use of an Internet referencing service to the operator of a search engine (like Google AdWords), in order to facilitate access to the trader's site by consumers domiciled in various Member States;
- the use of a top-level domain different from the one applicable in the trader's country;
- the mention of an international clientele composed of customers domiciled in various Member States, in order to establish that the trader was envisaging doing business with customers domiciled in the European Union, whatever the Member State.

4. MERELY INDICATIVE FACTS

4.1. CONCLUSION OF THE CONTRACT AT A DISTANCE

In the *Mühlleitner* case,[44] a consumer from Austria was connected to a German car seller through a German Internet search platform (mobile.de). She contacted the seller, using the telephone number given on the website, which included an international dialling code. Details of the vehicle in question were subsequently sent by e-mail. She was also informed that her Austrian nationality would not prevent her from acquiring a vehicle from the seller, taking immediate delivery of it. The CJEU held that 'both the establishment of contact at a distance, as in the present case, and the reservation of goods or services at a distance, or *a fortiori* the conclusion of a consumer contract at a distance, are indications that the contract is connected with such an [targeted] activity.'[45] It is not necessary, however, that the contract resulting from a targeted (online) activity is concluded at a distance.

4.2. CAUSATIVE CONNECTION

Furthermore, in the *Emrek* case the CJEU ruled that a causal link between the targeted (online) activity and the conclusion of the contract is not

---

[43] German Federal Court of Justice, 24.04.2013 (2013) *RIW* 563, marginal no. 22: '*Wij spreken Nederlands!*' on the website of a company that is based in Germany and is offering mobile homes for rent.
[44] Case C-190/11, *Mühlleitner v. A. Yusufi/W. Yusufi*, para. 45.
[45] *Ibid.*, para. 44.

required.⁴⁶ A consumer from Saarbrücken (Germany) heard from friends about a second-hand car dealer in Spicheren (France), a town close to the German border. He was not aware of the fact that the dealer maintained a website, which contained a German mobile telephone number. The consumer travelled to the premises of the dealer and bought a car. Later he brought an action against the dealer before a German court. The CJEU held that both the existence of a causal link and the fact that a trader uses a foreign telephone number in order to save his clients the cost of an international call can point to an intentionally 'directed activity' within the meaning of Article 17(1)(c) Brussels Ia Regulation.⁴⁷ The causative connection between the marketing activity and the conclusion of the consumer contract is, on the other hand, not a necessary element of the targeted activity test. In *Emrek* the Court pointed out convincingly that the consumer otherwise might encounter great difficulty to prove that he actually had looked up the trader's website before concluding the contract.⁴⁸

5. THE NEED FOR SPECIFIC CONFLICTS OF LAW RULES FOR INTERNATIONAL CONSUMER CONTRACTS IN THE DIGITAL SINGLE MARKET

As far as the underlying legal policy is concerned, there seems to be no need to adjust the existing conflict-of-law rules to the proposed directives. It is true that Article 6 Rome I refers to contracts concluded with parties regarded as being weaker, and therefore protects those parties by conflict of-law rules that are more favourable to their interests than the general rules.⁴⁹ But at the same time Article 6 Rome I Regulation balances the interest of the consumer to rely on the protective remedies foreseen by the law of the country where he has his habitual residence with the professional's interest to rely on the law of the country of its central administration/principal place of business (Article 19(1) Rome I Regulation). In fact, with its combination of the objective connecting factor of the consumer's habitual residence and the subjective connecting factor of the professional's intention to establish commercial relations with consumers from one or more other Member States (including that of the consumer's domicile),⁵⁰ the provision safeguards the legitimate interests of

---

46   Case C-218/12, *L. Emrek v. V. Sabranovic*, para. 32; CALLIESS, *supra* n. 9, Rome I Article 6, Consumer Contracts, marginal no. 54.
47   Case C-218/12, *L. Emrek v. V. Sabranovic*, paras. 29–30; as to the trader's intent as the key prerequisite in this regard see again Case C-144/09, *P. Pammer v. K. Schlüter*, para. 75.
48   Case C-218/12, *L. Emrek v. V. Sabranovic*.
49   Rome I, recital 23.
50   Case C-144/09, *P. Pammer v. K. Schlüter*, para. 75.

both parties to an international consumer contract. This approach is perfectly in line with the philosophy of the two proposed directives aiming at the creation of a digital single market to balance the interests of both parties to a contract.[51]

Things look different as far as the decision in a given case is concerned, whether or not the professional 'directed' its activity to the consumer country. As pointed out in the introduction (see above section 1), this decision is still crucial for a series of legal aspects in distance sales contracts.[52]

When applying the so-called targeted activity test, the competent court is required to make an overall assessment of the circumstances in which the consumer contract at issue was concluded. Article 6(1)(b) Rome I Regulation applies depending on the existence or absence of evidence mentioned on the non-exhaustive lists as established by the CJEU in the relevant case law.[53]

The problem with these lists is their lack of legal certainty[54] for professionals, especially as far as the role of the professional's website is concerned. It has in fact been criticised[55] that there are several inconsistencies. On the one hand, the CJEU makes clear that, since every website inherently has a worldwide reach, advertising on a website by a trader is in principle accessible in all states *irrespective of the intention* of the trader to target consumers outside the territory of the state in which it is established.[56] On the other hand, the Court held that the use of a top-level domain name other than that of the Member State in which the trader is established or the mention of telephone numbers with an international dialling code can be evidence capable of demonstrating the existence of an activity 'directed to' the Member State of the consumer's domicile.[57] According to these principles, nearly every website is directed to a global market because making use of top-level domains and/or phone number in this way is perfectly common even for small businesses in Europe. This is in stark contrast to the Court's statement that the use of online advertising methods as such does not pass the targeted activity test.

---

[51] Cf. the draft Directive on the online sale of goods, recital 19 and pp. 4, 14 of the Explanatory Memorandum; draft Directive on contracts for the supply of digital contents, recital 42 and p. 4 of the Explanatory Memorandum.

[52] Cf. once more the draft Directive on the online sale of goods, p. 7 of the Explanatory Memorandum: 'The proposal will not harmonise all aspects concerning contracts for the online and other distance sales of goods.'

[53] Case C-218/12, *L. Emrek v. V. Sabranovic*, para. 31.

[54] Legal certainty is what the commission aims at with its draft Directive on the online sale of goods, see recital 18.

[55] J. CLAUSNITZER, 'Anmerkung zur Entscheidung des EuGH – Zum Gerichtsstand bei Verbraucherverträgen via Internetangebot' (2011) *EuZW* 104–105 on Case C-144/09, *P. Pammer v. K. Schlüter*.

[56] Case C-144/09, *P. Pammer v. K. Schlüter*, para. 68.

[57] *Ibid.*, para. 83.

Generally, the factual elements listed by the Court are in no way binding for national courts.[58] This conflicts with the Commission's aim of promoting legal certainty for both businesses and consumers in the digital single market.[59] The proposals should be clear on this point. They should be accompanied by a detailed definition of the targeted activity test required by Article 6(1)(b) Rome I Regulation and Article 17(1)(c) Brussels I Regulation. This definition should be contained in these regulations, not in the proposed directives.[60]

As to the criteria of the targeted activity test, a professional who is offering his goods and services on his website should, in principle, be deemed to direct his activities to any country in the world. This presumption is justified by the mere fact that an Internet site is, by definition, accessible worldwide.[61] The professional should, however, be allowed to rebut this presumption with reference to the consumer's state of habitual residence by demonstrating:

- that on his website (a) he used different languages than the one(s) usually adopted in that state[62] and (b) he did not offer his goods and services in an international business language;
- that there is no causal link between the targeted (online) activity and the conclusion of the contract;[63] or
- that he took due diligence in ring-fencing his activities and that a certain consumer was addressed only accidentally by virtue of spill-over effects.[64]

---

[58] *Ibid.*, para. 93 (emphasis added): 'The following matters, *the list of which is not exhaustive, are capable of* constituting evidence from which it may be concluded that the trader's activity is directed to the Member State of the consumer's domicile, namely ... *It is for the national courts to ascertain whether such evidence exists*.'.

[59] Draft Directive on online sales, pp. 2, 3, 5 and 7 of the Explanatory Memorandum, and recitals 7, 9, 17, 18, 24, 26, 32, 34.

[60] On the highly problematic coexistence of a general rule (Article 6 Rome I) and a series of special rules see F. RAGNO, 'Article 6' in F. FERRARI (ed.), *Rome I Regulation. Pocket Commentary*, Sellier, Munich 2015, marginal no. 49 et seq.

[61] I do not share the position of recital 24 Rome I: anyone who uses a means of communication with a worldwide range (technically and by definition) *is* addressing the whole world. Any exceptions to this rule have to be demonstrated by the professional, see below.

[62] See Regulation (EU) 2016/679 of 27 April 2016 on the protection of natural persons with regard to the processing of personal data and on the free movement of such data, recital 23.

[63] Case C-218/12, *L. Emrek v. V. Sabranovic*, para. 32; CALLIESS, *supra* n. 9, Rome I Article 6, Consumer Contracts, marginal no. 54.

[64] CALLIESS, *supra* n. 9, Rome I Article 6, Consumer Contracts marginal no. 50, referring to the opinion of Attorney General Trstenjak in Joint Cases C-585/08 and C-144/09, *Alpenhof v. O. Heller / P. Pammer v. K. Schlüter,* para. 88 (on the risks of sending out spam); P. MANKOWSKI, *supra* n. 25, p. 474 et seq.

The targeted activity test is an objective test.[65] This follows from the binding nature of the consumer protection laws (Article 6(2) sentence 2 Rome I Regulation: 'provisions that cannot be derogated from'). It follows that a professional who is offering his goods and services on his website cannot avoid the application of the consumer protection laws of the consumer's state of habitual residence by unilaterally stating that he has no intention to conclude contracts with consumers with habitual residence in specific states ('disclaimer').

---

[65] Joint Cases C-585/08 and C-144/09, *Alpenhof v. O. Heller / P. Pammer v. K. Schlüter*, paras. 93 et seq.; cf. PLENDER and WILDERSPIN, *supra* n. 30, pp. 9–56 et seq.; CALLIESS, *supra* n. 9, Rome I Article 6, Consumer Contracts, marginal no. 50.

PART IV

NEW FEATURES OF STANDARD CONTRACTS IN THE DIGITAL MARKET

STANDARD TERMS AND TRANSPARENCY IN ONLINE CONTRACTS

Rodrigo Momberg

1. Introduction 189
2. Digital Content and Wrap Contracts 191
3. The Invisibility of Wrap Contracts 193
4. The Enforceability of Wrap Contracts 195
5. Transparency in EU Law 198
6. Transparency and Wrap Contracts 202
7. Curing Invisibility: Sufficient Notice and Specific Consent 204
   7.1. Sufficient Notice 204
   7.2. Specific Consent 205
8. Conclusions: The Unavoidable Assessment of Substantive (Un)Fairness . 206

1. INTRODUCTION

The idea of contract is based on consent: a party voluntarily enters into an agreement by which he imposes obligations on himself in (most of the time) exchange for something. Therefore, traditionally, a contract is defined as an agreement. Thus, the DCFR states that '[a] contract is an agreement which is intended to give rise to a binding legal relationship or to have some other legal effect'.[1] The proposals for a Directive on contracts for the supply of digital content and for online and other distance sales of goods also define a contract as 'an agreement intended to give rise to obligations or other legal effects'.[2] Then, in this view of contract, by the interplay of offer and acceptance, the parties bind themselves intentionally, creating a mutual relationship.

But even in a negotiated contract, which contract law doctrine usually presumes to be balanced, the agreement, or more precisely the consent of the

---
[1] Art. II.-1:101.
[2] Art. 2.7 of COM(2015) 634 final, 2015/0287 (COD) and 2(h) of COM(2015) 635 final, 2015/0288 (COD).

parties to the agreement, is far from perfect. Most of the time, the parties are not aware of or do not understand the terms of the contract beyond the core of the bargain, usually the price and the goods or services provided. Consent (or non-consent) has different levels. Thus, the ignorance of a party may refer to the existence of a contractual relationship, or the existence, content or meaning of some of the terms of the contract.[3]

Contract law recognises that a person can legally bind himself without knowing that he has done so, through his purported or hypothetical intention. Then, via a legal fiction, the law creates or supplements a contractual relationship. Today, the prevailing view on contract formation is the objective principle, which implies that the agreement must be understood from the external perspective of a reasonable observer, and not necessarily from the actual perception or understanding of the parties.[4] The reasons for this construction may be varied, but the main rationales seem to be the protection of the expectations of the other party, efficiency and legal certainty.[5]

The role (and existence) of consent has been particularly discussed with regard to standard form contracts. Thus, adhesion contracts have been critiqued on the basis that they lack real assent by the non-drafting party.[6] But the existence of a 'negotiated contract' to which both parties assent and understand the totality of its terms is no more than an idealisation. The mere existence of substantive default legal rules which supplement the consent of the parties makes any consent less complete and any contract more complex than the parties intended. A 'negotiated contract' usually means that the parties have bargained and discussed the core terms of the agreement, but secondary or ancillary terms, as well as gaps, are provided or filled by default contract law rules.

However, consent can also be considered not from a binary perspective (consent or non-consent) but also as a high-quality/low-quality continuum.[7] This means recognising that the level of consent is different in different contracts, and also for different terms in the same contract. A transaction between sophisticated

---

[3] See B. Bix, 'Consent in Contract Law' <http://web.law.columbia.edu/sites/default/files/microsites/law-theory-workshop/files/Bix%20Workshop%20Paper.pdf> accessed 27.05.2011.

[4] A classical formulation of the objective principle can be found in *Smith v. Hughes* (1871) LR 6 QB 597, where Blackburn J stated that: 'If, whatever a man's real intention may be, he so conducts himself that a reasonable man would believe he was assenting to the terms proposed by the other party, and that other party upon that belief enters into the contract with him, the man thus conducting himself would be equally bound as if he had intended to agree to the other party's terms.'

[5] See J. Steyn, 'Contract Law: Fulfilling the Reasonable Expectations of Honest Men' (1997) 113 *Law Quarterly Review* 433, stressing that the objective theory of contract promotes certainty and predictability in the resolution of contractual disputes.

[6] See generally, M.J. Radin, *Boilerplate: The Fine Print, Vanishing Rights, and the Rule of Law*, Princeton University Press, Princeton 2013.

[7] D. Barnhizer, 'Reassessing Assent-based Critiques of Adhesion Contracts' in L. DiMatteo and M. Hogg (eds.), *Comparative Contract Law: British and American Perspectives*, Oxford University Press, Oxford 2015.

parties with similar bargaining power is presumed to require a high-quality level of consent. There, the parties have strong incentives to examine and evaluate the terms of the bargain. On the other hand, B2C contracts of adhesion require a low-quality level of consent on the part of the consumer. The incentives present in complex transactions are absent in consumer contracts, and, as explained below, it is not rational to expect the consumer to give informed consent beyond the core terms of the contract.

On this view, the idealised-informed consent, that is, the ideal situation in which the parties consent to *and* understand all of the terms of the contract, cannot be considered as reason (and condition) for the enforceability of the contract. Low-quality consented contracts are also enforceable, but their enforceability requires the examination of further conditions which need to be assessed against the relative vulnerability of one of the parties, and also, as proposed here, against the context in which the transaction is concluded, that is, whether it is a traditional paper-based contract or an online contract. However, as stated below, even the existence of low-quality consent by the non-drafting party can be contended in the case of some wrap contracts.

But assuming that consent (or some kind of consent) exists in standard form contracts (paper or online) does not remove the problem of determining to which terms that consent has been given and which of those terms are enforceable by the provider. There are multiple approaches to this question, for instance those giving force only to the core terms, or to terms for which adequate notice has been given to the non-drafting party, or only to those considered reasonable. The case of online contracting is examined below.

2. DIGITAL CONTENT AND WRAP CONTRACTS

Online contracting is one of the main ways in which users can gain access to digital content.[8] Digital content is defined in the proposed Directive as data which is produced and supplied in digital form, services allowing the creation, processing or storage of data in digital form, and services allowing sharing of and any other interaction with data in digital form.[9]

The situation in which a party binds himself without being conscious of having done so is a common feature of the digital environment. Most online contracts, either for digital content or for other kinds of products or services, are entered into in the form of 'wrap contracts'. The concept of wrap contracts

---

[8] A note on terminology: the term 'consumer' will be avoided in most of the chapter, since the analysis is intended to cover also transactions on which one of the parties cannot be legally qualified as 'consumer'. User and non-drafting party will be used instead.

[9] Art. 2(1)(a) of the Proposal for a Directive on certain aspects concerning contracts for the supply of digital content, COM(2015) 634 final, 2015/0287 (COD).

is related to form and not to substance. Wrap contracts are adhesion contracts in which terms are presented in a non-traditional format and to which the non-drafting party assents in a non-traditional manner.[10] In this context, 'non-traditional' means different to paper contracts which are presented for the non-drafting party to sign in order to become binding.[11]

Wrap contracts are not limited to the digital environment. The concept originated in relation to the clear plastic wrapping around software packages. There, the contract (a *shrink-wrap* contract) is wrapped in that plastic and the opening of the package is meant to constitute assent to the conditions detailed in it. Similarly, a *box-wrap* contract involves the opening of a box as a sign of assent to the terms of the contract. In both cases, the contract is packaged *with* the product.

In the digital environment, the most common wrap contracts are *click-wrap* and *browse-wrap* agreements. In a click-wrap agreement, the non-drafting party must click on a dialogue box containing the words 'I agree' or similar in order to proceed with a transaction. The presentation of the terms may differ (for instance, they may appear in a scrollable box or simply at a hyperlink), but the essential element of click-wrap contracts is that they require a positive action by the non-drafting party (clicking the 'I agree' box), which can be interpreted as a manifestation of consent.[12]

Browse-wrap contracts do not directly present the terms of the contract to the non-drafting party, but rather the terms are accessible via one or more hyperlinks (labelled for instance 'Terms of use' or 'Legal terms'). The non-drafting party is not required to take any positive action, that is, is not required to access and agree to the terms by clicking on the link. The mere use of the website or downloading the digital content is considered a sufficient manifestation of consent.

The main difference between click-wrap and browse-wrap contracts is related to the attitude that is required by the non-drafting party. Whereas in a click-wrap contract that party must take a positive action (clicking on the 'I agree' box) in order to complete the transaction, in a browse-wrap contract the non-drafting party is not required to take any specific action with regard to its acceptance of the terms of the contract, which are simply made available by the provider via a hyperlink.

---

[10] See N. KIM, *Wrap Contracts. Foundations and Ramifications*, Oxford University Press, Oxford 2013, p. 2, who defines wrap contract as 'unilaterally imposed set of terms which the drafter purports to be legally binding and which is presented to the nondrafting party in a nontraditional format.'

[11] See L. TRAKMAN, 'The Boundaries of Contract Law in Cyberspace' (2008–2009) 38 *Pub. Cont. L.J.* 187, 188; KIM, *supra* n. 10, p. 3, 36–39.

[12] See KIM, *supra* n. 10, p. 39. Similarly, the author mentions 'tap-wrap' contracts, which pop up on mobile devices, requiring a tap of the finger on the screen as manifestation of assent.

As a consequence, the non-drafting party in a click-wrap contract cannot access the services without *doing something* (clicking the box), but in a browse-wrap contract he is not required to take any positive assenting action, and the services can be used without expressly agreeing to the terms available via the hyperlink or even without accessing to them. Therefore, it is more likely that the non-drafting party to a click-wrap contract may be aware that he is actually entering a contractual relationship; however, this would usually not be the case for parties to browse-wrap contracts.

3. THE INVISIBILITY OF WRAP CONTRACTS

These new methods of contracting create a number of challenges for the traditional understanding of a contract as the product of an offer and acceptance process and, in general, to the concept and consequences of contract as an agreement.

The fact that online users are commonly unaware of the terms of the legal relationship they enter into, and even of the fact that they are actually entering into a legal relationship with the provider at all, can be described as the *invisibility* of online or wrap contracts. This invisibility is the consequence of different features of online contracting.

First, wrap contracts are not subject to any formality, not even evidentiary formalities. There are no limitations on form, as in paper contracts, where physical constraints and costs encourage providers to avoid or to reduce the use of 'material' contracts. The inexistence of any form means that the non-drafting party can easily ignore the existence of a contract. For instance, in the case of browse-wrap contracts, the hyperlink 'Terms of use', intended to set out the terms and conditions of the contract the user is entering into, is normally placed at the bottom of a website, which means that the user is able to access the services of the provider without any knowledge of the existence of those terms. But even in the anomalous case that the user notes their existence, the user usually does not access the hyperlink or connect the expression 'Terms of use' with a legal document that can impose on him legal or contractual obligations or the waiver of his rights. Similarly, because the attention of online customers is mainly on the product or service they want to acquire, terms of 'contracts that are accessible only by clicking or scrolling can be easily dismissed by the user as unimportant if they are noticed at all', since the form and presentation of a contract, like its function, sets the parties' expectation.[13] For instance, signature witnessed by a public notary, required in some jurisdictions for the sale of land, is considered a manifestation of the importance of the transaction. Equally, customers will consider contracts that are only accessible via a hyperlink to

---

[13] KIM, *supra* n. 10, p. 178 and p. 200.

be objectively irrelevant. Reasonable people are unlikely to believe that merely surfing a website could constitute a contractual relationship.

Secondly, in the physical marketplace, users are normally not confronted with lengthy contracts (in the form of a document) for 'insignificant' transactions, services or activities. The situation should be no different for online transactions, and therefore users reasonably do not expect a hyperlink to contain a detailed and extensive regulation (in terms of rights and obligations) of the transactions they are interested in.

Thirdly, the form or presentation of online contracts also discourages the user from reading their terms or even noticing the existence of a contract. For instance, even when the 'I agree' button or box is immediately visible, access to the terms of the contract may require continuous and lengthy scrolling. Access to the terms interrupts the process of completing the transaction, for instance the downloading of digital content, which constitutes the user's main interest. Furthermore, legal notices and contract terms cannot compete with 'arresting visual images and even animation', which naturally attract the interest of the user.[14]

Fourth, online wrap contracts may be produced and modified by providers (for instance, adding new terms of use) at zero or almost zero cost when compared to paper contracts. Thus, there are no costs associated with printing or physical storage. More importantly, there are no costs in terms of customer goodwill, which could otherwise be affected by constantly having to deal physically with lengthy contracts for minor or insignificant transactions.[15]

Fifth, the fact that some services, for example email services, cloud services and social networks, seem to be provided free of charge may cause users to think that they have no legal or contractual relationship with the provider. The lack of any ostensible counter-performance by the user makes him unaware that he might be assuming obligations, or waiving or being deprived of rights.

It has been argued that even in the case of click-wrap contracts it is not evident to the non-drafting party that he is entering a contractual relationship. Online customers do not weigh the clicking on a box with the physical act of placing a signature equally, and it has been said that they 'have trained themselves to click the "I agree" button as quickly as possible'.[16]

Thus, paradoxically, online contracts have become so ubiquitous that they seem to be invisible to users; in this sense, 'consumers may manifest assent without being aware of what they are doing.'[17] The situation described raises some questions related to the very existence of a contract. The lack of visibility of the contract itself (not only of some of its terms), its form (for

---

[14] KIM, *supra* n. 10, p. 62. Similarly, it has been stated that '[t]he imperfectly rational consumer deals with complexity by ignoring it': O. BAR-GILL, *Seduction by Contract: Law, Economics and Psychology in Consumer Markets*, Oxford University Press, Oxford 2012, p. 83.
[15] KIM, *supra* n. 10, p. 58.
[16] BARNHIZER, *supra* n. 7, p. 177, adding that 'providers are well aware of this behaviour.'
[17] KIM, *supra* n. 10, p. 59.

instance, a browse-wrap contract) and the context of the transaction (for instance, merely surfing a website) may imply that a reasonable user does not intend to be bound by the contract because he is reasonably unaware of its existence. Empirical studies have confirmed the invisibility of online contracts.[18] The consequences of this invisibility for the enforceability of wrap contracts are discussed in the next section.

4. THE ENFORCEABILITY OF WRAP CONTRACTS

As a general rule, and considering consent not as a binary phenomenon (consent or non-consent) but as a continuum from high-quality to low-quality consented contracts, it can be affirmed that, in general, click-wrap contracts can be considered enforceable, and browse-wrap contracts not.[19]

This has been the general approach of the American case law. Thus, in *Specht v. Netscape Communications Corp.*,[20] an arbitration clause included in a browse-wrap contract was declared unenforceable. In that case, the browse-wrap link stated 'Please review and agree to the terms of the *Netscape SmartDownload software license agreement* before downloading and using software', but users were not required to click on that link as a condition of downloading Netscape's software. The Court decided that users were not bound by Netscape licence because they had not viewed the licence agreement and therefore they had not assented to the contract.[21] In other words, downloading software does not necessarily mean that the user agrees to terms that he is not reasonably aware of.[22]

However, it seems that is not merely the fact that the contract is proposed in the form of a browse-wrap contract which makes the terms unenforceable, but whether the user has been provided with sufficient notice of the existence of the terms, and therefore of the contract. In *Hines v. Overstock.com, Inc.*,[23] the completion of the purchase did not require an express action on the 'Terms and

---

[18] See references in KIM, *supra* n. 10, p. 160, mentioning, for instance, that only between 0.05 percent and 0.22 percent of users access online end-user licence agreements.

[19] See TRAKMAN, *supra* n. 11, p. 216, stating that: 'All others factors being constant, click-wrap agreements are more likely to be declared enforceable than browse-wrap contracts'; and KIM, *supra* n. 10, p. 41, adding that: 'Courts generally uphold clickwrap agreements, although many commentators find their "take-it-or-leave-it" nature troubling.'

[20] 150 F. Supp 2d 585 (S.D.N.Y.), aff'd, 306F.3d 17 (2d Cir. 2002).

[21] *Ibid.* at 592.

[22] But see *Register.com v. Verio, Inc.* (356 F.3d 393), where the Court decided that the continuous use of a website is capable to make aware a costumer of the terms of use presented in a browse-wrap form. See also TRAKMAN, *supra* n. 11, p. 218, noting that '[t]here is no magic in holding click-wrap contracts enforceable and browse-wrap contracts unenforceable', concluding that courts may decide the enforceability of terms and conditions on grounds beyond the difference between browse-wrap and click-wrap contracts.

[23] 668 F. Supp. 2d 362 (E.D.N.Y. 2009) aff'd 380 Fed. Appx. 22 (2nd Cir. June 3, 2010).

conditions' of the provider, which were available at a link placed at the bottom of the page, only visible after scrolling. The Court found that the terms were unenforceable, stressing that 'unlike in other cases where courts have upheld browsewrap agreements, the notice that "Entering this Site will constitute your acceptance of these Terms and Conditions," was only available *within* the Terms and Conditions.'[24] Similarly, in *In re Zappos.com, Inc., Customer Data Security Breach Litigation*,[25] the Court stated that '[w]here, as here, there is no evidence that plaintiffs had actual notice of the agreement, the validity of a browsewrap contract hinges on whether the website provides reasonable notice of the terms of the contract.'[26]

In a recent case, *Nguyen v. Barnes & Noble, Inc.*,[27] the purchaser of two touchpads in an online 'liquidation sale' was notified the next day that his order had been cancelled due to a high volume of orders. The dispute arose about the enforceability of an arbitration clause included in the 'Terms of use', which were available via a hyperlink displayed conspicuously at the bottom of every webpage in the online checkout process. The 'Terms of use' page included a provision stating that 'By visiting any area of the Barnes & Noble.com Site ... a user is deemed to have accepted the Terms of Use.' The Court decided that the user had no actual or constructive knowledge of the agreement since at no point in the transaction was he required to take any affirmative action to agree to the terms in order to complete the transaction.[28]

Issues of validity of contracts and the question of when a term is effectively included in a contract are not generally regulated at the level of EU law, and therefore it is not simple to assess the effectiveness of click-wrap and browse-wrap contracts at that level.[29] The E-Commerce Directive[30] sets out in Article 10(3) that '[c]ontract terms and general conditions provided to the recipient must be made available in a way that allows him to store and reproduce him.' The opportunity for that availability, i.e. when the terms must be provided to the consumer, is not specified. More precise is the Services Directive,[31] which in Article 22(4) states as a general rule that the provider must supply the information in good time before the conclusion of the contract, or, where there is no written contract, before the service is provided. That information includes

---

[24] *Ibid.* at 367, emphasis added.
[25] 3:12-CV-00325-RCJ, 2012 WL 4466660 (D. Nev. Sept 27, 2012).
[26] *Ibid.* at 3.
[27] No. 12-566628, 2014 WL 4056549 (9th Cir. Aug. 18, 2014).
[28] See also *Southwest Airlines v. BoardFirst, LLC*, 3:06-CV-0891-B, 2007 WL 4823761 (N.D. Tex. Sept. 12, 2007), where the Court stated at 5 that 'the validity of a browsewrap license turns on whether a website user has actual or constructive knowledge of a site's terms and conditions prior to using the site.'
[29] M. Loos, 'The Regulation of Digital Content Contracts in the Optional Instrument of contract Law' (2011) 6 *ERPL (European Review of Private Law)* 729, 736.
[30] Directive 2000/31/EC.
[31] Directive 2006/123/EC.

'the general conditions and clauses, if any, used by the provider' (Art. 22(1)(f)). Similarly, the Consumer Rights Directive (CRD) requires for distance contracts, the trader to provide to the consumer with relevant information before the consumer is bound by the contract (Art. 6).[32]

However, these information duties are not complemented with explicit remedies. The consequences of infringing these duties are not clear, in particular with regard to the enforceability against the user of contract terms that are not supplied by the provider in a timely manner.

Conversely, the failed CESL included rules on the valid inclusion of terms in B2C contracts concluded by electronic means. Thus, Article 24(3)(e) required the seller to make contract terms available before the other party makes or accepts an offer, and Article 23(4) added that they must be made available in alphabetical or other intelligible characters, and on a durable medium. Significantly, Article 70(1) provided that the seller can only invoke standard terms against the other party only if that party was either aware of the terms or if the seller took reasonable steps to draw the consumer attention to them, before or when the contract was concluded. Article 70(2) added that in B2C contracts terms are not considered to have been sufficiently brought to the consumer's attention by a mere reference to them in a contract document, even if the consumer signs the document.[33]

In the *Content Services* decision,[34] the ECJ ruled that, for the purposes of Article 5(1) of the repealed Directive 97/7/EC on distance contracts, 'a business practice consisting of making the information referred to in that provision accessible to the consumer only via a hyperlink on a website of the undertaking concerned does not meet the requirements of that provision, since that information is neither "given" by that undertaking nor "received" by the consumer, within the meaning of that provision, and a website such as that at issue in the main proceedings cannot be regarded as a "durable medium"'.[35] The Court stressed that '[i]n a process of transmission of information, it is not necessary for the recipient of the information to take any particular action. By contrast, where a link is sent to a consumer, he must act in order to acquaint himself with the information in question and he must, in any event, click on that link.'[36] With regard to the characteristics of a 'durable medium', the Court

---

[32] More references to similar provisions on other Directives are provided in paragraph 5 below.
[33] The direct source of the provision seems to be Art. II.-9:103 of the DCFR. A similar rule is provided by Art. 2:104 PECL.
[34] C-49/11 (05.07.2012).
[35] Paragraph 1 of Art. 5(1) of the repealed Directive 97/7/EC on distance contracts provided that: 'The consumer must receive written confirmation or confirmation in another durable medium available and accessible to him of the information referred to in Article 4 (1) (a) to (f), in good time during the performance of the contract, and at the latest at the time of delivery where goods not for delivery to third parties are concerned, unless the information has already been given to the consumer prior to conclusion of the contract in writing or on another durable medium available and accessible to him.'
[36] C-49/11 (05.07.2012) at 33.

indicated that such a medium must allow the consumer 'to store the information which has been addressed to him personally, ensures that its content is not altered and that the information is accessible for an adequate period, and gives consumers the possibility to reproduce it unchanged'.[37]

From the decision of the Court, it can be presumed that standard terms included in browse-wrap contracts are unlikely to be considered validly incorporated into the agreement. Even for click-wrap contracts the mere availability of the terms via a hyperlink could be insufficient to bind the consumer to the standard terms, but all the circumstances of the case should be assessed.[38] As has been correctly pointed out, '[d]epending on how the click-wrap licence is technically set up, the consumer's consent may be required either at the moment when the consumer when the contract is concluded, when the digital content is downloaded, or when the digital content is installed on the consumer's hardware, or a combination thereof.'[39] Then, standard terms presented to the consumer by a click-wrap only when the digital content is downloaded or installed, that is, *after* the conclusion of the contract, could be deemed ineffective.

In sum, the positive action that the non-drafting party is required to take with click-wrap contracts in order to complete the transaction can be considered a manifestation of his assent. Therefore, in principle, click-wrap agreements can be deemed enforceable, so long as the terms of the contract are presented to the consumer before the completion of the transaction and in accordance with the requirements established by the CJEU in *Content Services*. In addition, the assent of the non-drafting party should be placed on the lowest end of the consent-quality continuum, and therefore further assessment of which terms are enforceable are required, and further mechanisms for the protection of the non-drafting party provided. One of these mechanisms, but not the only one, is transparency.

5. TRANSPARENCY IN EU LAW

Under the principle of transparency, the provider must ensure that the other party has an opportunity to become aware of the terms of the contract before its

---

[37] Ibid. at 43. Art. 2(10) CRD defines durable medium as 'any instrument which enables the consumer or the trader to store information addressed personally to him in a way accessible for future reference for a period of time adequate for the purposes of the information and which allows the unchanged reproduction of the information stored.'

[38] The click-wrap contract of the case was a multiwrap, where the terms are presented to the user on a hyperlink situated alongside or below the 'I accept' box or icon. To access the terms, the user needs to click on the hyperlink, but that is not necessarily required as a previous step for the clicking on the 'accept' box.

[39] M. Loos, 'The Regulation of Digital Content B2C Contracts in CESL', Centre for the Study of European Contract Law Working Paper Series No. 2013–10.

conclusion and that those terms must be understandable for the average party.[40] Transparency has also been described as the formal element that must be assessed with regard to the unfairness of standard terms.[41] As explained below, the principle of transparency has been recognised not only by EU legislation but also by the case law of the ECJ.

In EU consumer law, the obligation of the trader to provide the standard terms to the consumer is derived from recital 20 of the Unfair Contract Terms Directive (UCTD),[42] which provides that 'the consumer should actually be given an opportunity to examine all the terms'. Other EU law instruments include this obligation in more precise terms and not necessarily restricted to consumer transactions, for example: Article 13(1) of the Commercial Agency Directive[43] ('Each party shall be entitled to receive from the other on request a signed written document setting out the terms of the agency contract including any terms subsequently agreed'); Articles 3 and 5 of the Distance Marketing of Financial Services Directive[44] (obligation of the supplier to communicate to the consumer all the contractual terms, conditions and certain information before the conclusion of the contract and in durable means); Article 22(1) of the Services Directive[45] (obligation of the providers to make certain information available to the recipient of the services); and Article 10(3) of the E-Commerce Directive[46] (contract terms and general conditions provided to the recipient must be made available in a way that allows him to store and reproduce them). Article 8(1) CRD requires that in the case of distance contracts the trader must give the information provided for in Article 6(1) or make that information available to the consumer.

As stated above, this aspect of transparency is not sufficient in online contracts. The mere opportunity to read or the sole availability of the terms do not adequately protect the interest of the non-drafting party. Therefore, the most important aspect of the transparency principle is that confers content to the information duties of the trader. Then, the mere provision of information is not sufficient if that information is not presented in a form adequate for the proper understanding of the non-drafting party.

The UCTD addresses this aspect in Article 5, stating that contract terms provided by the trader 'must always be drafted in plain, intelligible language'.

---

[40] M. Loos, 'Transparency of Standard Terms under the Unfair Contract Terms Directive and the Proposal for a Common European Sales Law' (2015) 2 *ERPL* 179, 180.
[41] H.W. Micklitz, 'Chapter 3. Unfair Terms in Consumer contracts' in N. Reich, H.W. Micklitz, P. Rott and K. Tonner, *European Consumer Law*, 2nd ed., Intersentia, Cambridge 2014, p. 142.
[42] Directive 93/13/ECC.
[43] Directive 86/653/EEC.
[44] Directive 2002/65/EC.
[45] Directive 2006/123/EC.
[46] Directive 2000/31/EC.

The application of the 'core terms' exception also requires those terms to be in in plain intelligible language (Art. 4(2)).

The CRD requires in Articles 5(1) and 6(1) that certain information must be provided to the consumer in 'a clear and comprehensible manner' before the contract is concluded. Article 6(7) of the same Directive provides that 'Member States may maintain or introduce in their national law language requirements regarding the contractual information, so as to ensure that such information is easily understood by the consumer.' As mentioned before, Article 8(1) requires that in the case of distance contracts the trader must give the information provided for in Article 6(1) or make that information available to the consumer in plain and intelligible language and in a way that is appropriate to the means of distance communication used. Article 8(2) adds that if the consumer is under an obligation to pay, 'the trader shall make the consumer aware in a clear and prominent manner, and directly before the consumer places his order', of some of that information. Article 8(2) also states that a contract in which the trader fails to make the consumer aware that the order implies an obligation to pay is non-binding.

Accordingly, Article 10(1) of the E-Commerce Directive states that, for contracts concluded by electronic means, certain information must be given by the service provider 'clearly, comprehensibly and unambiguously and prior to the order being placed by the recipient of the service'. Importantly for the purposes of this contribution, Article 10(1)(a) includes 'the different technical steps to follow to conclude the contract' as part of the information that must be provided. Again, Article 22(4) of the Services Directive states that 'Member States shall ensure that the information which a provider must supply in accordance with this Chapter is made available or communicated in a clear and unambiguous manner, and in good time before conclusion of the contract or, where there is no written contract, before the service is provided'.

As mentioned above, the ECJ has also stressed the importance of transparency for the adequate protection of the consumer. In *Commission v. the Netherlands* the Court clearly stated that when implementing the UCTD Member States must expressly include this principle in their legislation as recognised in Article 5 of the Directive.[47]

---

[47] C-144/99 (10.05.2001) at 17, 18. It is interesting to note that at para. 20, the ECJ expressly relies on the reasoning of the Advocate General, who in his Opinion, when referring to the obligation of transparency of the trader by Art. 5 states that 'a provision which expressly reflects the principle [of transparency] laid down in the Directive is efficient and immediate in its effect, especially – as again the Commission observes – if it is borne in mind that the approach chosen by the Community legislature is specifically intended to require the seller or supplier to make sure at the outset that the contractual terms are plain and intelligible, thus ensuring that, before entering into the contract, the consumer has access to all the information needed to arrive at his decision in full knowledge of the facts' (para. 31 of Advocate's General Opinion).

In *Kásler*[48] the Court specified that the transparency requirement of plain and intelligible language of Article 5 UCTD 'cannot therefore be reduced merely to their being formally and grammatically intelligible.'[49] The Court added that the position of weakness of the consumer means that the requirement of transparency must be understood in a broad sense. Therefore, the average consumer must not only be aware of the existence of the terms, but also be able to assess the economic and legal consequences of the application and enforcement of the terms.[50] In other words, consumers should be able to predict when a particular event might trigger the application of a contractual term, and the effects of that application.

Recital 20 and Article 5 of the UCTD were interpreted by the Court in *Invitel*[51] in the sense that 'in the assessment of the "unfair" nature of a term, within the meaning of Article 3 of the Directive, the possibility for the consumer to foresee, on the basis of clear, intelligible criteria, the amendments, by a seller or supplier, of the GBC [general business conditions] with regard to the fees connected to the service to be provided is of fundamental importance.'[52] The same reasoning was applied in *RWE*,[53] where the Court added that the requirement of transparency 'is not satisfied by the mere reference, in the general terms and conditions, to a legislative or regulatory act determining the rights and obligations of the parties. It is essential that the consumer is informed by the seller or supplier of the content of the provisions concerned' and that '[t]he lack of information on the point before the contract is concluded cannot, in principle, be compensated for by the mere fact that consumers will, during the performance of the contract, be informed in good time of a variation of the charges and of their right to terminate the contract if they do not wish to accept the variation'.[54]

It seems clear that, for the Court, transparency involves more than the mere communication and availability of the information. The key aspect is that the average consumer (and in general the average party) is able to understand and use that information in order to foresee the effects of the contract and therefore decide whether or not to agree with its terms.

The failed CESL also included relevant provisions on the subject. Article 70 (Duty to raise awareness of not individually negotiated contract terms) provided in paragraph 1 that contract terms supplied by one party and not individually

---

[48] C-26/13 (30.04.2014).
[49] *Ibid.*, para. 71.
[50] *Ibid.*, paras. 73–75, and E. MACDONALD, 'Inequality of Bargaining Power and "Cure" by Information Requirement' in L. DIMATTEO and M. HOGG (eds.), *Comparative Contract Law: British and American Perspectives*, Oxford University Press, Oxford 2015, p. 164.
[51] C-472/10 (26.04. 2012).
[52] *Ibid.*, para 28.
[53] C-92/11 (21.03.2013).
[54] *Ibid.*, paras 50 and 55.

negotiated 'may be invoked against the other party only if the other party was aware of them, or if the party supplying them took reasonable steps to draw the other party's attention to them, before or when the contract was concluded.' Interestingly, paragraph 2 of the same article added that 'in relations between a trader and a consumer contract terms are not sufficiently brought to the consumer's attention by a mere reference to them in a contract document, even if the consumer signs the document.' Article 82 (Duty of transparency in contract terms not individually negotiated) expressly recognised the duty of transparency in contracts between a trader and a consumer. The provision stated that where a trader supplies contract terms which have not been individually negotiated with the consumer 'it has a duty to ensure that they are drafted and communicated in plain, intelligible language.'

The duty of transparency has been also been recognised in European instruments of contract law, such as the PECL (Art. 2:104) and the DCFR (Arts. II.-9:402; II.-3:105; II.-9:103).

Even when there is no express sanction for breach of the transparency principle, it has been argued that the ECJ case law suggests that a breach of transparency should be considered a presumption of unfairness.[55] Article 83(2)(a) CESL expressly provided this consequence for contracts between a trader and a consumer but Article 70 stated that the terms in question may not be invoked, so the contract is treated as having been made without those terms.

In this matter it should be remembered that under Article 4(2) UCTD the application of the 'core terms' exception also requires those terms to be in in plain intelligible language. This can be considered the first adverse consequence of a breach of transparency. In addition, some guidance can be sought in the provisions of the CRD related to distance contracts. Article 8(2)(2) considers the contract not binding on the consumer if the trader does not comply with the form conditions required to make the consumer aware that the order he is placing entails an obligation to pay. In this sense, a broad interpretation could mean that terms that do not comply with the transparency principle are not binding for the non-drafting party. This is a more protective and efficient solution than its consideration only as an element of unfairness.

6. TRANSPARENCY AND WRAP CONTRACTS

The disclosure of information is one of the main devices of consumer protection. The conventional assumption is that an informed consumer will make a better choice. If the consumer does not like the terms of the agreement (of which is he fully previously informed) he is free to go the market and choose another provider with better terms. Providers discarded by consumers would seek to

---

[55] Loos, *supra* n. 40, pp. 179, 184.

improve their terms and then the quality (in the sense of terms more favourable for consumers) of standard terms available for consumers will improve.

This assumption has been falsified on several grounds by behavioural studies, which emphasise the limitations of consumer choice by bounded rationality and cognitive limitations.[56]

The failures of the model are accentuated in online contracting. Transactions related to digital content are usually made in the form of wrap contracts that amplify the inefficiency of traditional disclosure and notice methods. The invisibility of wrap contracts have been addressed above (section 3), but it is useful to return to some of those ideas.

Thus, in paper contracts, all terms of the transaction are at some point visible to the non-drafting party *before* he signs the contract. The fact that most consumers do not read the contract which is handed over for signature does not change this. In fact, most consumers will read some of the clauses related to the core terms of the contract, which are usually found in the first part of the document. Cross-referencing or references to other documents are normally deemed invalid by consumer protection law. But cross-referencing is a common practice in online contracting, as online agreements typically incorporate hyperlinked terms by reference, that is, terms that only will be visible to the user by clicking on the hyperlink. There hyperlink terms may in turn incorporate other hyperlink terms. Sometimes, the hyperlink requires the user to scroll before it is visible. All of this has the consequence that the size of online contracts and the amount of terms that they include are larger than in paper contracts. This painful process of finding the information is likely to discourage rational consumers from reading the standard terms presented by the provider.

Furthermore, on digital contracts, most of the terms will be *available* for the non-drafting party, but not necessarily *visible*. The notice itself may be available but not easily visible, for instance if it is placed at the bottom of a webpage, where the user will normally only see it after scrolling, or hidden between other irrelevant information. In some cases, the recipient may receive notice after undertaking acts that constitute acceptance, for instance only when the digital content is downloaded or installed.

These characteristics create new challenges for the transparency principle. The mere communication of information, represented by the opportunity to examine the terms of the contract by accessing an hyperlink at which they are available, or by scrolling through a small pop-up window, seem to be clearly insufficient to comply with the requirements that contract provisions, and

---

[56] See MACDONALD, *supra* n. 50, p. 165, and R. HILLMAN, 'Online Consumer Standard Form Contracting Practices: A Survey and Discussion of Legal Implications' in J.K WINN (ed.), *Consumer Protection in the Age of the 'Information Economy'*, Ashgate, Abingdon 2013, stressing at p. 296 that 'many or most e-consumers may still have ample rational reasons not to read (e.g. lack of bargaining power, commonality of terms within and industry) and cognitive shortcomings that impede reading (e.g. information overload).'

information in general, must be provided in 'plain intelligible language', or 'clear and comprehensible manner' in order to ensure that such information is 'easily understood' by the non-drafting party.

The special characteristics of online contracts require a reinforcement of the conditions required to comply with transparency. Some of those measures are examined below.

7. CURING INVISIBILITY: SUFFICIENT NOTICE AND SPECIFIC CONSENT

7.1. SUFFICIENT NOTICE

In order to adequately protect the non-drafting party and for the construction of, at least, his presumptive consent, it is not sufficient that the *notice* of the existence of the terms is visible to that party: the *terms* themselves must be conspicuous and easily accessible, as in the case of paper contracts.[57]

This implies that the non-drafting party should be aware that he is entering into a contractual relationship. Before the completion of the transaction, the main terms of the contract should be conspicuously presented to the user. Particular prominence should also be given to terms that put the user in a situation of disadvantage or that deprive him of legal rights. Following the approach of the English Court of Appeal in *Office of Fair Trading v. Abbey National plc and others*,[58] the attention of the user should be drawn to the issues that usually are not on his focus when entering into the transaction, for example unilateral modification of the agreement, privacy or applicable jurisdiction.

As a counterpart of the non-drafting party's 'duty to read' under American law, it has been suggested that providers should have a 'duty to draft reasonably' imposed on them, by which drafting parties must make standard terms clear and readable, linking the presentation of the terms to their significance, stressing that the test should be 'how a reasonable person have communicated the mandatory nature of these terms?'[59] This would mean improving the visibility of terms, changing the terminology (e.g. from 'Terms of Use' to 'Legal Terms' or 'Contractual Obligations'), avoiding having to click a hyperlink as the only way to review the standard terms, and providing a clear means of assenting to or rejecting the terms.[60]

---

[57] On the contrary some courts in the USA have found that is sufficient that the notice itself was visible and accessible. See KIM, *supra* n. 10, p. 129.
[58] [2009] EWCA Civ 116, [2009] 1 All ER (Comm) 1097.
[59] KIM, *supra* n. 10, p. 183.
[60] *Ibid.*, pp. 176–192.

The simplification of the terms offered to the consumer would also improve the quality of the information and therefore transparency. Thus, clauses which not have any impact in the obligations of the parties could be eliminated and presented to the consumer in a separate form. For instance, mere instructions for use, terms related to the technical characteristics of the goods or services, or statements about the policy governing the service should be extracted from the 'Terms and conditions' and placed under a different heading. This would increase the visibility of the important terms.[61]

In sum, sufficient notice therefore should imply that the non-drafting party has to be able to use and process the information provided, or at least have a real opportunity to do so.

7.2. SPECIFIC CONSENT

Sufficient notice only resolves part of the problems related to online contracting. The user's knowledge that he is entering into a legal relationship is not always enough to protect his interests and expectations. It can be presumed that with adequate notice the non-drafting party will be aware of the core terms of the agreement. But the existence or content of other terms may still remain obscure.

When those terms have the effect of depriving the non-drafting party of some of his legal rights, imposing obligations or derogating from non-mandatory legal rules, it has been suggested that specific consent to be required from the non-drafting party.

Distinguishing between the primary transaction and ancillary or secondary terms, it has been suggested that in online contracts those terms should be enforceable only if the non-drafting party has expressly consented to them. This proposal is based on the assumption that users are, at the most, aware (and therefore have consented) to the core terms of the transaction, but not to ancillary or secondary terms that do not form part of the bargain.[62] Requiring specific consent would improve the quality of the agreement for the user, reflecting more accurately the benefits and costs of the transaction as a whole. Additionally, it would protect the user from undesired terms.

Articles 8(2)(2) and 22 of the CRD are examples of this technique, imposing specific consent to the obligation to pay for an order by the consumer, and to any extra payment in addition to the remuneration agreed upon for the trader's main contractual obligation.[63]

---

61 *Ibid.*, pp. 195–196.
62 *Ibid.*, pp. 192–200.
63 For the problems on the interpretation and implementation of this provision, see A. DE FRANCESCHI, 'The EU Digital Single Market Strategy in Light of the Consumer Rights Directive: The "Button Solution" for the Internet Cost Traps and the Need for a More Systematic Approach' (2015) 4 *EuCML* 144–148.

However, this alternative has also been criticised. It has been argued that providers would be able to direct the preferences of users placing higher prices to terms he considers undesirable, discouraging their selection in favour of other offered for free or at insignificant cost. In addition, non-professional users would face serious problems in measuring the value of most of the terms offered by the provider.[64]

8. CONCLUSIONS: THE UNAVOIDABLE ASSESSMENT OF SUBSTANTIVE (UN)FAIRNESS

This contribution has intended to highlight the difficulties and problems that online contracting creates for the law regulating standard form contracts, in particular with regard to the principle of transparency as recognised in EU law. Even when the proposed Directives do not address this question, the issue can be addressed by resorting to principles and criteria already present in other instruments of EU law and in the case law of the ECJ. In this sense, transparency, as the most perfect expression of the information paradigm in consumer protection law, should be reinforced when applied to online contracts. However, transparency alone cannot completely cure inequality. The law should be wary of increasing procedural fairness at the cost of reducing control over substantive fairness.[65]

Procedural requirements can be useful if they are combined with substantive assessment. In EU law this should not be difficult; transparency can be complemented by the fairness test of Article 3 UCTD.[66] However, the precise scope of the 'core exemption' of Article 4 UCTD can be problematic. If the exemption is broadly constructed by the courts, more terms will be excluded from the fairness test. An approach similar to the English Court of Appeal in *Office of Fair Trading v. Abbey National* seems adequate: confine 'core terms' to those embodying the 'essential bargain', that is, those which the consumer will normally or reasonably considers the object of his consent, exercising effectively his freedom of contract with regard to them.[67] Protection should then be ensured with regard to terms that a consumer will not focus on when entering the bargain.

---

[64] BARNHIZER, *supra* n. 7, p. 180, adding at p. 182 that most of the apathy of consumers with regard to terms out of the core bargain is 'because the terms have little value to the consumer at the time of contracting.'
[65] In this sense, see HILLMAN, *supra* n. 56, p. 296, arguing that 'by increasing the fairness of the process of contracting, the law may legitimse some terms that courts otherwise would be likely to strike on unconscionability or other grounds.'
[66] Similarly, BARNHIZER, *supra* n. 7, p. 176.
[67] MACDONALD, *supra* n. 50, p. 160.

Accordingly, the fairness test should especially consider the existence of consideration or counter-performance for the obligations assumed or the rights waived by the consumer. Because of the particular characteristics of online contracts, it can be assumed that most of the ancillary terms included by the trader do not provide any valuable counter-performance for the user. In other words, they 'do not form part of the consideration for the primary transaction'.[68]

What should be avoided is the legitimisation of unfair terms through the provider's compliance with procedural rules, or through a selective-assent system of enforceability.[69] Transparency should benefit the consumer, not damage his position through a formalistic legitimisation of unfair terms. In the case of contracts where one of the parties is in a position of structural imbalance, both because of the relative position of the parties *and* the context of the transaction, such as the consumer in digital contracting, procedural fairness should not prevail over substantive fairness.

If in the circumstances of the case, which in online contracting necessarily include qualifying the contract offered as click-wrap or browse-wrap, it is found that a contract has been concluded, transparency can be useful to determine which terms (if any) of those proposed by the provider have been incorporated into the contract. This stage involves an examination of procedural unconscionability or formal fairness. But even if it is decided that the terms have been effectively incorporated into the contract, their substantive fairness still needs to be assessed.

Therefore, a reinforced and adapted principle of transparency should be the first barrier for standard terms in online contracts, but should never mean that some terms will become legitimate because they are fully transparent. A necessary second step should always involve an assessment of substantive fairness (Art. 3 UCTD) and consider in particular the position of the parties and the existence of counter-performance. In sum, even where the low-quality consent of the non-drafting party in online contracts may be cured through sufficient notice, specific consent and a reinforcement of transparency, adequate protection of users in the online environment makes a substantive assessment unavoidable.

---

[68] BARNHIZER, *supra* n. 7, p. 179.
[69] In the same sense, BARNHIZER *supra* n. 7, p. 183: '[a]ssent should not be the focus for critique or debate in online adhesion contracting. Assent doctrines do not provide a firm basis for arguing the illegitimacy of contracts of adhesion. Even if form producers invested in developing mechanisms aimed at producing high-quality assent, the level of assent from such mechanisms would likely not increase in a substantial way'.

CONTRACTS CONCLUDED BY ELECTRONIC MEANS IN CROSS-BORDER TRANSACTIONS

'Click-Wrapping' and Choice-of-Court Agreements in online B2B Contracts

Martin GEBAUER

1. Introduction 209
2. Normative Background 211
   2.1. The Purpose Beyond the Wording of Article 25(1) Brussels I *bis* Regulation 211
   2.2. Communication by Electronic Means According to Article 25(2) of the Brussels I *bis* Regulation 212
3. The European Court of Justice and Choice-of-Court Agreements Concluded by Electronic Means 213
   3.1. Dispute in the Main Proceedings 214
   3.2. The Reasoning of the ECJ 215
4. Jurisdictional Consequences of the Decision Given by the ECJ in Case C-322/14 217
   4.1. B2B and Consumer Contracts 217
   4.2. Durable Records 218
   4.3. Choice-of-Court Agreements as Compared to Choice-of-Law Agreements 218

1. INTRODUCTION

The European Commission has identified the completion of the Digital Single Market (DSM) as one of its main political priorities. The Commission's Website states that the

> 'internet and digital technologies are transforming our world … It's time to make the EU's single market fit for the digital age – tearing down regulatory walls and moving

from 28 national markets to a single one. This could contribute € 415 billion per year to our economy and create hundreds of thousands of new jobs.'[1]

That sounds quite good. And again one is deeply impressed by the Commission's quantification skills: it is not €400 or €480 billion but rather €415 billion per year that moving to a single market could contribute to the European economy. However, according to the Commission's so-called *Digital Single Market Factsheet*, 'obstacles remain to unlock these potentials'.[2] In reality, cross-border online services seem to represent only a small percentage of overall EU online services. The Commission's website with reference to its priorities points out:[3]

'But at present, markets are largely domestic in terms of online services. Only 7% of EU small- and medium-sized businesses sell cross-border. This needs to change – putting the single market online.'

My focus in this article is on such cross-border contracts and specifically on what might be perceived as a barrier, a regulatory wall, or an obstacle, namely formal requirements in contracts concluded online.

There is little doubt that the way we conclude contracts today has dramatically changed as compared to 25 years ago. In the digital world, the formal requirements of a contract may seem to be a throwback to the old world, the non-digital world. By formal requirements I do not mean those specific 'formal requirements' for consumer contracts concluded by electronic means which are meant to enable consumers to be informed about all the costs before entering into a binding contract. In that regard, Article 8(2) of the Consumer Rights Directive[4] introduced a so-called 'button solution' in order to provide sanctions for 'internet cost traps'.[5] Nor do I mean domestic contracts entered into by parties based in one jurisdiction. I mean cross-border contracts, the kind of contracts which the Commission has in mind when it complains about the largely domestic nature of markets in the context of the digital single market. But what is special about cross-border contracts in terms of formal requirements?

In cross-border transactions, well-informed parties ought to anticipate *inter alia* two different sets of issues that may arise if they are facing legal problems. The first issue is about jurisdiction to adjudicate: which court will be competent

---

[1] <http://ec.europa.eu/priorities/digital-single-market_en> accessed 28.05.2016.
[2] <https://ec.europa.eu/priorities/sites/beta-political/files/dsm-factsheet_en.pdf> accessed 28.05.2016.
[3] <http://ec.europa.eu/priorities/digital-single-market_en> accessed 28.05.2016.
[4] Directive 2011/83/EU of the European Parliament and of the Council of 25 October 2011 on Consumer Rights [2011] OJ L304/64.
[5] See A. De Franceschi, 'The EU Digital Single Market Strategy in Light of the Consumer Rights Directive: The "Button Solution" for Internet Cost Traps and the Need for a More Systematic Approach' (2015) *Journal of European Consumer and Market Law* 144, 146 et seq., pointing out the fragmentary implementation in the Member States.

to decide a legal dispute? The second issue is about choice of law: which law will govern the dispute between the parties?

In order to address these two issues in advance, well-informed parties should consider introducing into their contract a forum selection clause and a choice-of-law clause. At least in contracts between businesses or professionals, there is quite a lot of autonomy left to the parties regarding both the determination of the competent court and the applicable law. However, in terms of formal requirements, there is quite a big difference between forum selection and choice-of-law clauses. European Union law only addresses the formal requirements of forum selection clauses, and this is the reason I want to focus on the formal requirements of these procedural contracts, on choice-of-court agreements.

In the first part of this chapter, I will set out the normative background of the formal requirements of forum selection clauses in European Union law. In the second part, I will focus on the first decision that the European Court of Justice gave last year regarding the formal requirements of a choice-of-court agreement concluded by electronic means. And finally, in the third part, I will present some of the consequences that the Court's decision will probably have on European choice-of-court agreements.

2. NORMATIVE BACKGROUND

2.1. THE PURPOSE BEYOND THE WORDING OF ARTICLE 25(1) BRUSSELS I *BIS* REGULATION

Starting with my first part, the normative background in terms of the formal requirements of jurisdiction clauses, according to Article 25(1) of the Brussels I *bis* Regulation:

> '[t]he agreement conferring jurisdiction shall be either:
> (a) in writing or evidenced in writing;
> (b) in a form which accords with practices which the parties have established between themselves; or
> (c) in international trade or commerce, in a form which accords with a usage of which the parties are or ought to have been aware and which in such trade or commerce is widely known to, and regularly observed by, parties to contracts of the type involved in the particular trade or commerce concerned.'

In the context of contracts concluded by electronic means, I want to leave out the *practices* established between the parties as well as international trade *usages*, and rather concentrate here on the form of *writing*. However, it is very important to see that the structure and purpose of all these formal requirements goes beyond the general purposes we have in mind when we think of formal requirements. The 'purpose of the formal requirements imposed by Article [25] is to ensure that the consensus between the parties is in fact established.'[6] This is what the ECJ already stated forty years ago in 1976 in the *Colzani* case,[7] regarding the old Brussels Convention, but the court's reasoning is still valid today under the Brussels I *bis* Regulation. This purpose is especially important with regard to the incorporation of those general terms and conditions containing a choice-of-court agreement into a contract. Instead of leaving the question as to whether the general terms and conditions have been incorporated into the contract to be resolved under the applicable national substantive rules, the Brussels approach is to ensure consensus between the parties by means of imposing formal requirements.[8] These formal requirements have to be interpreted autonomously, taking into account the purpose of ensuring a material consensus between the parties which lies behind the European formal requirements on choice-of-court agreements.

2.2. COMMUNICATION BY ELECTRONIC MEANS ACCORDING TO ARTICLE 25(2) OF THE BRUSSELS I *BIS* REGULATION

During the 1990s it became obvious that the new forms of communication would fundamentally change the way in which contracts are concluded. As Fausto Pocar wrote in his explanatory report to the Lugano Convention, electronic commerce should not be obstructed by inappropriate formal requirements.[9] For this reason, about fifteen years ago the European legislator introduced a second paragraph to Article 23 of the Brussels I Regulation, which is today Article 25(2) of the Brussels I *bis* Regulation. It states as follows:

> 'Any communication by electronic means which provides a durable record of the agreement shall be equivalent to "writing".'

The concept 'equivalent to "writing"' allows electronic communication to serve as a substitute for an agreement written on paper. In order to understand the purpose of the rule on electronic communication, it might be helpful to make reference to the Explanatory Memorandum on the Proposal for a Council Regulation, presented by the Commission in Brussels on 14 July 1999, 17 years ago.[10] The amendment

---

[6] Case 24/76, *Colzani v. RÜWA* [1976] ECR 1831, 1832 (first sentence of the decision's summary). The original text refers to Article 17 of the Brussels Convention.
[7] *Ibid.*
[8] MAGNUS/MANKOWSKI/MAGNUS, *Brussels Ibis Regulation*, 2nd ed. 2016, Art. 25, paras. 89–90.
[9] Explanatory Report by FAUSTO POCAR to the Convention on jurisdiction and the recognition and enforcement of judgments in civil and commercial matters, signed in Lugano on 30 October 2007 [2007] OJ C319/29, para. 109.
[10] COM(1999) 348 final.

'takes account of the development of new communication techniques. The need for an agreement "in writing or evidenced in writing" should not invalidate a choice-of-forum clause concluded in a form that is not written on paper but accessible on screen. The reference, of course, is mainly to clauses in contracts concluded by electronic means.'[11]

What the European legislator certainly had in mind by the end of the last century was the exchange of emails.[12] However, the rule was drafted in an open way in order to cover future technical developments.[13] The criterion of there being at least a 'durable record' of the agreement aims to make sure that any clause conferring jurisdiction on a certain court, and at the same time excluding jurisdiction of all other courts, was in fact the subject of consensus between the parties.

One has to bear in mind that a valid choice-of-court agreement has far-reaching consequences. The derogative effect of an exclusive choice-of-court agreement means that access to justice is precluded in any other jurisdiction, even if the default rules of the Brussels I *bis* Regulation, so those applicable in the absence of an agreement, would open a basis of jurisdiction in that place.

For this reason, certainly not every message on a website or other electronic medium will meet the formal condition of providing a durable record. Voicemails, video conferences and other communication not in writing do not seem to be covered by the rule; the same is true for SMS messages on mobile phones unless they can be reproduced on paper.[14]

3. THE EUROPEAN COURT OF JUSTICE AND CHOICE-OF-COURT AGREEMENTS CONCLUDED BY ELECTRONIC MEANS

For the very first time, the ECJ ruled on choice-of-court agreements concluded by electronic means in May 2015 in its decision in *CarsOnTheWeb*.[15] The following question was referred to the Court by a German Regional Court:

---

[11] COM(1999) 348 final.
[12] Kropholler/von Hein, *Europäisches Zivilprozessrecht*, 9th ed., 2011, Art. 23, para. 41; Reithmann/Martiny/Hausmann, *Internationales Vertragsrecht*, 8th ed., 2015, para. 8.73, with further references to national court decisions.
[13] Rauscher/Mankowski, *Europäisches Zivilprozess- und Kollisionsrecht*, Volume I, 4th ed., 2016, Art. 25 Brüssel Ia-VO para. 126.
[14] Magnus/Mankowski/Magnus, *supra* n. 8, Art. 25 para. 131; Kropholler/von Hein, *supra* n. 12, Art. 23, para. 41; P. Gottwald in *Münchener Kommentar zur ZPO*, 4th ed., 2013, Art. 23 EuGVO para. 46; Reithmann/Martiny/Hausmann, *supra* n. 12, para. 8.73.
[15] Case C-322/14, *Jaouad El Majdoub v. CarsOnTheWeb.Deutschland GmbH* (2015) NJW (Neue Juristische Wochenschrift) 2117.

'Does so-called "click wrapping" fulfil the requirements relating to a communication by electronic means within the meaning of Article 23(2) [of the Brussels I Regulation]?'[16]

What is 'click-wrapping'? Well, for a start, probably every reader has entered into a contract which is click-wrapped, even if you have never heard the term before. So called 'click-wrap agreements' require the user to declare that they consent to general terms and conditions that do not open automatically on the website but only open if the user clicks a certain icon saying something like: 'Click here to view the terms and conditions'. Typically, the user must then confirm consent by clicking another box or icon saying something like: 'OK', 'I agree' or 'I accept'.

The ECJ held

'that the method of accepting the general terms and conditions of a contract for sale by "click-wrapping", concluded by electronic means, which contains an agreement conferring jurisdiction, constitutes a communication by electronic means which provides a durable record of the agreement, within the meaning of that provision, where that method makes it possible to print and save the text of those terms and conditions before the conclusion of the contract.'[17]

3.1. DISPUTE IN THE MAIN PROCEEDINGS

Before I come to the arguments of the ECJ, I would like to outline the factual background of the underlying case. The claimant in the main proceedings, a car dealer established in Cologne (Germany), purchased from the website of the defendant an electric car for a very good price. The defendant's registered office was in Germany as well. However, the sale was cancelled by the seller, the later defendant. The seller alleged that the car was damaged and that for this reason he had cancelled the sale. The claimant did not believe that and suggested that the reason given was only a pretext for the cancellation of the sale, which was disadvantageous to the seller because of the low sale price.

The claimant brought an action seeking the transfer of ownership of the car before the German court. By contrast, the defendant in the main proceedings contended that the German courts did not have jurisdiction in the case. Article 7 of the defendant's general terms and conditions for internet sales transactions, accessible on the defendant's website, contained an agreement conferring jurisdiction on a court in Leuven (Belgium) where the parent company of the defendant was registered.

---

[16] *Ibid.*, para. 19.
[17] *Ibid.*, para. 40.

The claimant in the main proceedings took the view that the jurisdiction clause in the defendant's general terms and conditions had not been validly incorporated into the sales agreement, as it was not in writing in accordance with the requirements in Article 23(1)(a) of the Brussels I Regulation. The requirements of Article 23(2) of the Brussels I Regulation were met only if the window containing those general terms and conditions opened automatically. It was submitted that a click-wrapping system was not sufficient to incorporate general terms and conditions. Moreover, the defendant submitted that the agreement conferring jurisdiction in the click-wrapped terms and conditions was also invalid because it was arbitrary and unexpected.

3.2. THE REASONING OF THE ECJ

The reasoning of the ECJ is quite interesting as it combines literal, historical, teleological and even systematic arguments. The third chamber of the Court, composed of five judges, decided, after hearing the Advocate General, to proceed to judgment without an Opinion of the Advocate General. Presumably the result of the judgment seemed to be quite clear to the Court.[18]

The Court started with its often repeated argument that the requirements laid down by Article 23 of the Brussels I Regulation must be strictly interpreted in so far as they exclude both jurisdiction as determined by the general principle of jurisdiction, being the courts of the state in which the defendant is domiciled, and the special additional rules of jurisdiction.[19] Then the ECJ went on to emphasise that any clause conferring jurisdiction upon a specific court must in fact have been the subject of consensus between the parties. In the present case, this seemed to be clear to the ECJ, which held that the 'purchaser expressly accepted the general terms and conditions at issue, by clicking the relevant box on the seller's website.'[20]

Furthermore, the Court reasoned based on a *literal* interpretation of the phrase 'provides a durable record' that the provision requires there to be only the 'possibility' of providing a durable record of the agreement conferring jurisdiction, regardless of whether the text of the general terms and conditions has actually been durably recorded by the purchaser before or after he clicks the box indicating that he accepts those conditions.[21]

Historical and teleological arguments are taken by the Court from the already mentioned Explanatory Memorandum on the Proposal for a Council Regulation,

---

18   W. WURMNEST (2015) *EuZW (Europäische Zeitschrift für Wirtschaftsrecht)* 565, 567.
19   Case C-322/14, *Jaouad El Majdoub v. CarsOnTheWeb.Deutschland GmbH* (2015) *NJW* 2117, para. 25.
20   *Ibid.*, para. 31.
21   *Ibid.*, para. 33.

in which the Commission stated in 1999 that a choice-of-forum clause concluded in a form that is not written on paper but accessible on screen should not be invalidated.[22] The main purpose of the amended jurisdiction rule for the ECJ is 'to simplify the conclusion of contracts by electronic means.'[23]

And finally, there are two systematic arguments brought forward by the ECJ. The first one refers to the Explanatory Report by Fausto Pocar to the Lugano Convention, signed in 2007.[24] This report also clarifies that the 'test of whether the formal requirement … is met is … whether it is possible to create a durable record of an electronic communication by printing it out or saving it to a backup tape or disk or storing it in some other way.' And the report goes on to say that the rule 'merely indicates that electronic communication is considered to be in writing "if it provides a durable record", even if no such durable record has actually been made.'[25]

The second systematic argument refers to and distinguishes from another judgment of the ECJ, where the Court held in 2012 with reference to Article 5 of the Directive on the Protection of Consumers in respect of Distance Contracts,[26] that a business practice consisting of making information accessible only via a hyperlink on a website did not meet the requirements of receiving 'written confirmation or confirmation in another durable medium' according to that Directive.[27] However, according to the ECJ, this interpretation cannot be applied to the Brussels I Regulation, since both the wording of Article 5(1) of Directive 97/7,[28] which expressly requires the communication of information to consumers in a durable medium, and the objective of that provision, which is specifically consumer protection, differ from those of Article 23 of the Brussels I Regulation.[29]

---

[22] COM(1999) 348 final.
[23] Case C-322/14, *Jaouad El Majdoub v. CarsOnTheWeb.Deutschland GmbH* (2015) NJW 2117, paras. 35–36.
[24] Explanatory Report by FAUSTO POCAR to the Convention on jurisdiction and the recognition and enforcement of judgments in civil and commercial matters, signed in Lugano on 30 October 2007 [2007] OJ C319/29, para. 109.
[25] *Ibid.*
[26] Directive 97/7/EC of the European Parliament and of the Council of 20 May 1997 on the protection of consumers in respect of distance contracts [1997] OJ L144/19. See today Article 8(7) Directive 2011/83/EU of the European Parliament and of the Council of 25 October 2011 on consumer rights [2011] OJ L304/64.
[27] Case C-49/11, *Content Services Ltd v. Bundesärztekammer*, para. 51.
[28] Compare Article 8(7) Directive 2011/83/EU of the European Parliament and of the Council of 25 October 2011 on consumer rights [2011] OJ L304/64.
[29] Case C-322/14, *Jaouad El Majdoub v. CarsOnTheWeb.Deutschland GmbH* (2015) NJW 2117, paras. 37 et seq.

4. JURISDICTIONAL CONSEQUENCES OF THE DECISION GIVEN BY THE ECJ IN CASE C-322/14

What are, in the end, the main consequences to be drawn from the click-wrapping decision of the ECJ?

4.1. B2B AND CONSUMER CONTRACTS

The first consequence is very important to the whole rationality of the decision: Consumer rights in terms of questions of jurisdiction do not seem to be affected at all. As the ECJ pointed out, the objective of Article 23 Brussels I Regulation (Art. 25 Brussels I *bis* Regulation) is not the protection of consumers. For a valid choice-of-court agreement in a consumer contract to stand, the very strict requirements of the Brussels regime set out in Articles 17 and 19 of Brussels I *bis* Regulation have to be fulfilled. To put it in simple words, according to these provisions choice-of-court agreements are only valid if they are designed in favour of the consumer.

However, according to Article 19(1) of Brussels I *bis* Regulation, the jurisdiction provisions in favour of the consumer may – *inter alia* – be departed from by an agreement 'which is entered into after the dispute has arisen'. Since Article 25 Brussels I *bis* Regulation addresses all choice-of-court agreements, including those entered into by a consumer, a click-wrap agreement could also be valid, within the limits of Article 19 Brussels I *bis*, to the consumer's disadvantage if entered into after the dispute has arisen. That said, it is quite unlikely that any consumer would willingly click through the opposing party's website in order to agree to a new forum selection clause.

As a consequence, the click-wrapping decision will mainly have effects on contracts concluded between businesses. And in that regard, it seems to be a sound decision. Certainly it is in line with the Development of the Digital Single Market and it confirms a practice which is already established.[30] The consent declared by a click combined with the possibility, required by the ECJ, 'to print and save the text of those terms and conditions before the conclusion of the contract'[31] makes it more comparable to an exchange of emails than to an oral agreement.[32]

---

[30] P. MANKOWSKI (2015) *LMK (Lindenmeier Möhrung Kommentierte BGH-Rechtsprechung)* 369738; W. WURMNEST, *supra* n. 18, pp. 565, 567, 568.

[31] Case C-322/14, *Jaouad El Majdoub v. CarsOnTheWeb.Deutschland GmbH* (2015) NJW 2117, para 40.

[32] W. WURMNEST, *supra* n. 18, p. 565, 567, 568. For a different view see Amtsgericht Geldern, 20.04.2011 (2011) *NJW-RR (Neue Juristische Wochenschrift – Rechtsprechungsreport)* 1503; P. MANKOWSKI, *supra* n. 30, p. 369738; RAUSCHER/MANKOWSKI, *supra* n. 13, Art. 25 Brüssel Ia-VO, para. 129.

4.2. DURABLE RECORDS

Article 25(2) Brussels I *bis* Regulation states that:

> '[a]ny communication by electronic means which provides a durable record of the agreement shall be equivalent to "writing".'

As above, the prevalent view is that voicemails, video conferences and SMS messages on mobile phones are not covered by the rule unless they can be reproduced on paper.[33] There may be some changes in the interpretation of the rule in the future. Voicemails and video conferences certainly provide a 'durable record'. However, for now it seems appropriate to exclude them from the scope of the rule. As an 'equivalent to "writing"', the communication should be readable without the need to transcribe it into written form. The ECJ was right in requiring the possibility 'to print and save the text of those terms and conditions before the conclusion of the contract'.[34]

4.3. CHOICE-OF-COURT AGREEMENTS AS COMPARED TO CHOICE-OF-LAW AGREEMENTS

Contracts containing a choice-of-court agreement will usually contain a choice-of-law clause as well. It is interesting to see how differently the European legislator addresses these types of contract. As noted above, the formal requirements of the Brussels regime go beyond the general purposes we have in mind when we think of formal requirements in substantive law. The 'purpose of the formal requirements imposed by Article [25][35] is to ensure that the consensus between the parties is in fact established.'[36] Under the regime of the Rome I Regulation, on the other hand, consent and the formal requirements of a choice-of-law agreement are left to the applicable national law of the *lex causae* which may even be the law of a third non-EU state (Arts. 3(5), 10 and 11 Rome I Regulation).

Imagine that the general terms and conditions contain not only a choice-of-court but also a choice-of-law agreement, to be opened within a click-wrapping

---

[33] MAGNUS/MANKOWSKI/MAGNUS, *supra* n. 8, Art. 25, para. 131; KROPHOLLER/VON HEIN, *supra* n. 12, Art. 23, para. 41; P. GOTTWALD, *supra* n. 14, Art. 23 EuGVO, para. 46; REITHMANN/MARTINY/HAUSMANN, *supra* n. 12, para. 8.73.
[34] Case C-322/14, *Jaouad El Majdoub v CarsOnTheWeb.Deutschland GmbH*, para. 40.
[35] The original text refers to Article 17 of the Brussels Convention.
[36] Case 24/76, *Colzani v. RÜWA* [1976] ECR 1831, 1832 (first sentence of the decision's summary). For the European autonomous concept of agreement and consensus see KROPHOLLER/VON HEIN, *supra* n. 12, Art. 23, paras. 23 et seqq.; M. GEBAUER, 'Das Prorogationsstatut im Europäischen Zivilprozessrecht' in H. KRONKE and K. THORN (eds.), *Festschrift für Bernd von Hoffmann zum 70. Geburtstag*, 2011, pp. 577–588, at 578–582.

system. Here at first glance we seem to be outside the autonomous determination of the European Union law. According to Article 3(5) Rome I Regulation, '[t]he existence and validity of the consent of the parties as to the choice of the applicable law shall be determined in accordance with the provisions of Articles 10, 11 and 13.' And according to Article 10(1), '[t]he existence and validity of a contract, or of any term of a contract, shall be determined by the law which would govern it under this Regulation if the contract or term were valid.'

Under the national substantive law referred to in Articles 3(5), 10 and 11 of the Rome I Regulation, the question of the validity of a choice-of-law clause contained within a click-wrapping system may be regarded as a matter of consent rather than as a matter of form. National substantive laws differ quite a lot even within the EU in terms of the valid inclusion of general terms and conditions into a contract.[37] As a consequence, in the example of a combined choice-of-court/choice-of-law agreement, the first may be valid according to the standards developed by the ECJ under Article 23 of the Brussels I Regulation (Art. 25 Brussels I *bis* Regulation), while the second may be invalid according to the standards of some national legal systems, applicable in terms of Articles 3(5) and 10(1) of the Rome I Regulation. Of course, the main contract has to be separated from the choice-of-law agreement and the latter is not identical with the choice-of-court agreement. However, this situation is not satisfactory in a European system of private international law.

An alternative approach could be to localise the problem not within the 'validity' of a contract or its single terms under Article 3(5) of the Rome I Regulation, but rather within Article 3(1) of the Rome I Regulation and the question as to whether the choice between the parties has been made 'expressly or [is] clearly demonstrated by the terms of the contract'. As a consequence, an autonomous interpretation could be developed, similar to the test put forward by the ECJ in terms of Article 23 of the Brussels I Regulation (Article 25 Brussels I *bis*). The incorporation of standard terms containing a choice-of-law clause can be regarded as a matter of a choice 'made expressly' within Article 3(1) of the Rome I Regulation.[38] Instead of determining whether a term has been incorporated into a contract and the transparency of the terms according to divergent national law, European standards could be developed, at least to a certain extent, for choice-of-law clauses contained in general terms and conditions.[39]

---

[37] See, for example, K. ZWEIGERT and H. KÖTZ, *An Introduction to Comparative Law*, translated by T. WEIR, 3rd ed. 1998, pp. 333–347; F. RANIERI, *Europäisches Obligationenrecht*, 3rd ed., 2009, pp. 323–403; H. KÖTZ, *Europäisches Vertragsrecht*, 2nd ed., 2015, pp. 191–216.

[38] M. MCPARLAND, *The Rome I Regulation on the Law Applicable to Contractual Obligations*, 2015, para. 9.26.

[39] Compare RAUSCHER/VON HEIN, *Europäisches Zivilprozess- und Kollisionsrecht*, Volume III, 4th ed., 2016, Art. 3 Rom I-VO, para. 43; Landgericht München I, 19 April 2011, Die deutsche Rechtsprechung auf dem Gebiete des Internationalen Privatrechts im Jahre 2011 (IPRspr. 2011), para. 25 (p. 53); regarding the so-called battle of the forms: Nomos Kommentar/ LEIBLE, *Rom-Verordnungen*, 2nd ed., 2015, Art. 3 Rom I-VO, para. 73.

Generally speaking, in the interests of the uniform application of party autonomy within the European instruments of private international law, it is preferable, where this seems possible, to develop European rules to address the basic requirements of an express choice-of-law agreement.[40] Following the click-wrapping decision rendered by the ECJ, it is suggested that standards analogous to those now applying to jurisdiction clauses should also apply to choice-of-law agreements in cross-border online contracts.

---

[40] E. JAYME, 'Kodifikation und Allgemeiner Teil im IPR' in S. LEIBLE and H. UNBERATH (eds.), *Brauchen wir eine Rom 0-VO?*, 2012, pp. 33, 38–40.

PART V

ONLINE PLATFORMS IN THE 'SHARING ECONOMY'

CROWDSOURCING CONSUMER CONFIDENCE

How to Regulate Online Rating and Review Systems in the Collaborative Economy

Christoph Busch

1. Introduction 223
2. More Reputation, Less Regulation? 225
   2.1. Reputation and Trust in the Collaborative Economy............ 225
   2.2. Reputational Enforcement as a Substitute for Regulation? 227
3. Recent Regulatory Initiatives 229
   3.1. National Level... 229
   3.2. European Level.. 230
4. Key Elements of a Regulatory Framework for Reputation Systems 232
   4.1. A European Standard for Online Reputation Systems........... 233
   4.2. Transparency and Consumer Information 234
   4.3. Submission of Reviews....................................... 235
   4.4. Processing of Reviews.. 237
   4.5. Publication of Reviews....................................... 239
   4.6. Consolidated Ratings.. 240
   4.7. Right of Reply 242
5. Conclusion... 242

1. INTRODUCTION

The rise of online platforms which facilitate peer-to-peer transactions has dramatically changed the digital economy over the last decade. New platform business models have emerged in various sectors from retail and travel, to transportation and banking. Many of the platforms are essentially 'matchmakers' that connect one group of customers with another group of customers.[1] Ride-sharing apps such as Uber, Lyft or BlaBlaCar bring together drivers and riders; websites like Airbnb or Wimdu connect people with spare rooms with those who are looking for a place to stay; and EatWith helps food lovers to find home chefs for private dinners.

The enabling factor that is essential for the functioning of such collaborative economy platforms is *trust*. A key role in building the necessary trust among strangers using a collaborative economy marketplace is played by so-called reputational feedback systems based on qualitative reviews and numerical ratings tied to the profile of a platform user. In order to promote trust, it is necessary that the reviews and ratings provided on the platform are trustworthy and free from bias or manipulation. Fake reviews or other forms of manipulation can undermine the trust in the reputation system and the business model of the platform itself. Recently regulators in several Member States have tackled this issue and started drafting rules for online reputation systems. While the establishment of quality criteria for rating and review mechanisms can support the functioning of the platform markets and help to avoid market failures, it is clear that differing national rules for reputation systems may create barriers in the European single market.

The key argument of this contribution is that there is a case for harmonisation of the rules for online rating and review systems at the European level in order to create a level playing field for the collaborative economy. Following the model of the 'new approach' which has been efficient and successful in the area of product safety, a flexible and innovative regulatory strategy could be a combination of an EU Directive, which defines 'essential requirements' for trustworthy and fair reputational feedback systems, and a harmonised European service standard which spells out the legal and technical details.

The chapter is structured as follows. Section 2 explains how reputation systems can contribute to the development of generalised trust among users of sharing economy platforms and thus complement existing regulatory tools for enhancing consumer confidence. Section 3 gives an overview of current regulatory initiatives regarding online ratings and reviews at national and EU level. Section 4 presents the idea of a European standard for reputation systems and outlines key elements of a future European regulatory framework.

---

[1] D.S. EVANS and R. SCHMALENSEE, *Matchmakers: The New Economics of Multisided Platforms*, Harvard Business Review Press, Boston 2016; see also G.G. PARKER, M.W. VAN ALYSTNE and S.G. CHOUDARY, *Platform Revolution*, Norton, New York 2016.

2. MORE REPUTATION, LESS REGULATION?

2.1. REPUTATION AND TRUST IN THE COLLABORATIVE ECONOMY

Online platforms like Airbnb, EatWith or Uber are often referred to as part of the so-called 'collaborative economy'.[2] Although there is no universally accepted definition of the 'collaborative economy',[3] which is also referred to as the 'share economy'[4] or 'peer-to-peer economy',[5] it may be conceptualised as 'any marketplace that brings together distributed networks of individuals to share or to exchange otherwise underutilized assets'.[6] From a (consumer) contract law perspective it is important that collaborative economy platforms are mainly used for transactions between private individuals on an occasional basis, but increasingly also by micro-entrepreneurs and small businesses. As a result the established line between consumer and business is blurred. Moreover, the rise of the collaborative economy increases the number of 'triangular' relationships involving a seller or service provider, a customer and the operator of the platform. From a similar perspective, economists refer to such platforms as two-sided (or multi-sided) markets.[7] A characteristic feature of multi-sided markets is that what the platform operator sells to the participants on one side of the market is access to the participants on the other side of market and vice versa.

No market can exist without trust. In close-knit communities of a small scale the trust required for commercial exchanges is mainly built through personal contact and repeated interactions between individuals that create 'reputation bonds'.[8] Trust is also the 'social glue' that facilitates collaborative consumption and the functioning of collaborative economy platforms. Without the development of trust among the users of a multi-sided platform, people

---

[2] See the European Commission Communication, 'A European agenda for the collaborative economy', COM(2016) 320 final.

[3] R. BOTSMAN and R. ROGERS, *What's mine is yours: The rise of collaborative consumption*, 2010; see also the contributions in R. SCHULZE and D. STAUDENMAYER (eds.), *Digital Revolution: Challenges for Contract Law in Practice*, Nomos, Baden-Baden 2015.

[4] See e.g. C. MELLER-HANNICH, 'Share Economy and Consumer Protection' in R. SCHULZE and D. STAUDENMAYER (eds.), *Digital Revolution: Challenges for Contract Law in Practice*, Nomos, Baden-Baden 2015, p. 119.

[5] See e.g. M. COHEN and A. SUNDARARAJAN, 'Self-Regulation and Innovation in the Peer-to-Peer Sharing Economy' (2015) 82 *The University of Chicago Law Review Dialogue* 116.

[6] C. KOOPMAN, M. MITCHELL and A. THIERER, 'The Sharing Economy and Consumer Protection Regulation: The Case for Policy Change' (2015) 8 *Journal of Business, Entrepreneurship & the Law* 529, 531.

[7] See the seminal work by J.C. ROCHET and J. TIROLE, 'Platform Competition in Two-Sided Markets' (2003) *Journal of the European Economic Association* 990.

[8] On the role of reputation bonds as a means of private ordering, see L. BERNSTEIN, 'Opting Out of the Legal System: Extralegal Contractual Relations in the Diamond Industry' (1992) 21 *Journal of Legal Studies* 115–157; see also R.C. ELLICKSON, *Order Without Law: How Neighbors Settle Disputes*, Harvard University Press, Cambridge MA 1991.

would not be willing to let strangers stay in their apartments or take them along for a ride. From this perspective, it is no coincidence that Airbnb describes itself as a 'trusted community marketplace'.[9] However, online marketplaces with geographically dispersed buyers and sellers who do not engage in repeated transactions are as such not very conducive to the traditional mechanisms for building trust. In order to create the trust among strangers that is a necessary requirement for the functioning of collaborative consumption marketplaces, almost all online platforms use some sort of reputational feedback system, i.e. some sort of record of qualitative reviews and numerical ratings tied to the profile of a platform user. Some platforms even encourage users to link their profiles to their online presence on other social media platforms (e.g. Facebook, Twitter, LinkedIn) in order to add additional layers of reputational certification.[10] In summary, it is probably no exaggeration to say that reputation systems are the very heart of many online platforms and that the collaborative economy can be referred to as a 'reputation economy'.[11]

In a certain sense online platforms emulate the trust-building mechanisms of small-scale communities.[12] In the offline world, reputational information travelled haphazardly through word of mouth, gossip and rumour.[13] Today, the rating and review systems of collaborative economy platform provide a technological solution for 'electronic word of mouth'. The ratings and reviews provide information about the transactional behaviour of users. Opportunistic behaviour is revealed and the information is disseminated among other market participants.[14] As a consequence, information asymmetries among market participants are reduced. Sellers, service providers or customers who have demonstrated that they are not trustworthy are punished via negative reviews and are eventually ostracised. Indeed, empirical research indicates that favourable online reviews have a direct positive impact on the revenue generated by sellers.[15] In other words, positive or negative ratings and reviews often translate to high, low or no sales respectively.[16] In short, reputation

---

[9] <https://www.airbnb.com/about/about-us?locale=en> accessed 15.06.2016.
[10] <https://www.airbnb.com/help/article/198/what-is-the-social-connections-feature?locale=en> accessed 15.06.2016.
[11] See H. Masum and M. Tovey (eds.), *The Reputation Society*, MIT Press, Cambridge MA 2011; see also J. Blocher, 'Reputation as Property in Virtual Economies' (2009) 118 *Yale Law Journal*, Pocket Part 120 <http://thepocketpart.org/ 2009/01/19/blocher.html.> accessed 15.06.2016.
[12] O. Lobel, 'The Law of the Platform' (2016) *Minnesota Law Review* (forthcoming).
[13] On the importance of gossip for social control see Ellickson, *supra* n. 8, p. 133.
[14] T. Dietz, *Global Order Beyond Law: How Information and Communication Technologies Facilitate Relational Contracting in International Trade*, Hart, Oxford 2014, p. 225.
[15] M. Luca, 'Reviews, Reputation, and Revenue: The Case of Yelp.Com', Harvard Business School NOM Unit Working Paper No. 12–016 (15 March 2016), available at SSRN: <http://ssrn.com/abstract=1928601> accessed 15.06.2016.
[16] K. Finley, *Trust in the Sharing Economy: An Explorative Study*, University of Warwick 2013, p. 18.

systems serve as a way of social control or community policing by performing an information transmission, verification and monitoring function that incentivises good behaviour.[17]

2.2. REPUTATIONAL ENFORCEMENT AS A SUBSTITUTE FOR REGULATION?

From a regulatory perspective, reputation systems fulfil a role similar to more traditional means of market regulation (e.g. mandatory disclosure rules). Some authors therefore claim that reputation systems are a way of creating 'self-policing communities' which make traditional forms of consumer regulation superfluous.[18] They argue that 'reputational enforcement' of decent market behaviour via ratings and reviews is superior to traditional instruments of consumer law in combatting market problems such as information asymmetries. This view resonates with the critics of 'disclosurism' who believe that mandated disclosure as a regulatory instrument should be abandoned and information should be left to markets which voluntarily provide advice to consumers in the form of ratings, rankings or reviews.[19] From a different perspective, reputational feedback mechanisms have been conceptualised as a new form of consumer empowerment leading to 'more informed and empowered consumers'.[20]

However, reputation systems fulfil their trust-building role only if the reputation scores they produce faithfully represent and predict future quality and behaviour. Unfortunately, the system design of many reputation mechanisms used by online platforms is less than perfect, which makes it relatively easy to manipulate ratings and reviews. Indeed, there seems to be a serious quality problem as recent reports about fake reviews on online shopping websites like Amazon.com indicate.[21] A recent report by the UK Competition &

---

[17] J.Y. LEE, 'Trust and Social Commerce' (2015) 77 *University of Pittsburgh Law Review* 137, 166.
[18] For such a rather optimistic view see KOOPMAN, MITCHELL and THIERER, *supra* n. 6; O. LOBEL, *supra* n. 12; see also A. THIERER, C. KOOPMAN, A. HOBSON and C. KUIPER, 'How the Internet, the Sharing Economy, and Reputational Feedback Mechanisms Solve the "Lemons Problem"', Mercatus Working Paper (26 May 2015), available at SSRN: <http://ssrn.com/abstract=2610255>, accessed 15.06.2016.
[19] O. BEN-SHAHAR and C.E. SCHNEIDER, *More than you wanted to know: The failure of mandated disclosure*, Princeton University Press, Princeton 2014, 185; see also O. BEN-SHAHAR and C.E. SCHNEIDER, 'Coping with the failure of mandated disclosure' (2015) 11 *Jerusalem Review of Legal Studies* 83–93. For an approach based on personalised disclosures see C. BUSCH, 'The Future of Pre-Contractual Information Duties: From Behavioural Insights to Big Data' in C. TWIGG-FLESNER (ed.), *Research Handbook on EU Consumer and Contract Law*, Edward Elgar Publishing, Cheltenham 2016 (forthcoming), available at SSRN: <http://ssrn.com/abstract=2728315> accessed 15.06.2016.
[20] KOOPMAN, MITCHELL and THIERER *supra* n. 6, p. 541.
[21] 'Amazon sues 1000 "fake reviewers"', *The Guardian* (18 October 2015) <www.theguardian.com/technology/2015/oct/18/amazon-sues-1000-fake-reviewers> accessed 15.06.2016; see

Market Authority also found ample evidence of fake reviews, negative reviews not being published and businesses paying for endorsements without this being made clear to consumers.[22] Moreover, there are indications that the design of reputation systems can inadvertently promote discriminatory behaviour.[23]

Reputation systems also suffer from some inherent weaknesses. Systemic problems which have been highlighted in the economic literature include herding behaviour, whereby earlier reviews influence the evaluation by subsequent reviewers, strategic reciprocity, where reviewers fear retaliatory negative reviews on platforms that allow reciprocal reviewing, and self-selection, where consumers who are more likely to be satisfied with a product are also more likely to purchase and review it.[24] Empirical studies on existing review platforms also show that many reputation systems generate implausible distributions of star-ratings that are unlikely to reflect true product quality. An analysis of ratings on Airbnb found that nearly 95% of properties listed on the platform show an average user-generated rating of either 4.5 or 5 out of 5 stars; virtually none have less than a 3.5-star rating.[25]

In summary, the analysis reveals a mixed picture. Considering the weaknesses and vulnerabilities of online reputation systems, it is probably an exaggeration to say that ratings and reviews can be a complete substitute for traditional means of market regulation. Total delegation of regulatory functions to the platforms is not feasible. Nevertheless, it remains true that a well-functioning reputation system can complement and reinforce existing market regulation that governs the activities of online platforms, such as rules on mandatory disclosures.[26] This complementary function is particularly important considering the fact that transaction amounts on online platforms are often rather small and parties are located in different jurisdictions. Legal recourse is therefore in many cases either impractical or unavailable.[27] The situation is aggravated by the fact that the protective rules of consumer law, which are drafted for B2C transactions, are usually not applicable for C2C transactions that are typical for the collaborative economy.[28]

---

also F MAROTTA-WURGLER, 'Even More Than You Wanted to Know About the Failures of Disclosure' (2015) 11 *Jerusalem Review of Legal Studies* 63, 71.

[22] Competition & Market Authority, 'Online reviews and endorsements – Report on the CMA's call for information' (19 June 2015); see also House of Lords, 'Online Platforms and the Digital Single Market' (20 April 2016), paras. 292–299.

[23] V. KATZ, 'Regulating the Sharing Economy' (2015) 30 *Berkeley Tech. L.J.* 1067, 1118.

[24] For an overview see G. ZERVAS, D. PROSERPIO and J.A. BYERS 'A First Look at Online Reputation on Airbnb, Where Every Stay is Above Average', Working Paper (28 January 2015) <http://ssrn.com/abstract=2554500> accessed 15.06.2016.

[25] *Ibid.*, p. 1.

[26] KATZ, *supra* n. 23, p. 1117.

[27] J.Y. LEE, 'Trust and Social Commerce' (2015) 77 *University of Pittsburgh Law Review* 137, 140.

[28] See C. BUSCH, H. SCHULTE-NÖLKE, A. WIEWIÓROWOWSKA-DOMAGALSKA and F. ZOLL, 'The Rise of the Platform Economy: A New Challenge for EU Consumer Law?' (2016) 5 *EuCML (Journal of European Consumer and Market Law)* 3, 4.

3. RECENT REGULATORY INITIATIVES

3.1. NATIONAL LEVEL

Growing concerns about the integrity of reputation mechanisms have recently prompted regulatory initiatives in a number of EU Member States. The spectrum ranges from traditional instruments such as guidelines issued by national market watchdogs and legislative amendments to consumer laws to rather novel tools such as standards drafted by national standardisation bodies.[29] In July 2013 the French Association for Standardization (AFNOR) published a standard that provides general principles and detailed requirements for the collection, screening and publication of online consumer reviews.[30] In May 2015 the Danish Consumer Ombudsman published guidelines on the publication of user reviews setting up requirements pursuant to the Danish Marketing Practices Act.[31] In May 2016 the French Senate adopted the legislative proposal of the *Loi pour la République numérique* which will amend the *Code de la consommation* in order to define transparency requirements and quality standards for rating and review systems.[32] The variety of different initiatives provides a fine example of regulatory pluralism in the European Union.[33]

These regulatory texts are complemented by a growing body of case law from Member State courts and national competition and consumer protection

---

[29] On service standards as regulatory tools see C. Busch, 'DIN-Normen für Dienstleistungen – Das europäische Normungskomitee (CEN) produziert Musterverträge' [2010] *NJW (Neue Juristische Wochenschrift)* 3061; see also P. Delimatsis, 'Standard-Setting in Services – New Frontiers in Rule-Making and the Role of the EU', TILEC Discussion Paper 2015–013 (June 2015) <http://ssrn.com/abstract=2616618> accessed 15.06.2016.

[30] AFNOR, French Standard NF Z 74–501 – Avis en ligne de consommateurs – Principes et exigencies portant sur les processus de collecte, moderation et restitution des avis en ligne de consommateurs (19 July 2013).

[31] Danish Consumer Ombudsman, *Guidelines on publication of user reviews* (1 May 2015), <www.consumerombudsman.dk/Regulatory-framework/dcoguides/Guidelines-on-publication-of-user-reviews> accessed 15.06.2016.

[32] Projet de loi pour une République numérique, Assemblée nationale, projet de loi, n° 3318; see also L. Grynbaum, 'Loyauté des plateformes: un champ d'application à redéfinir dans les limites du droit européen, A propos du projet de loi pour une République numérique', *La Semaine Juridique, Edition Générale*, no. 16, 18 April 2016, p. 778.

[33] See C. Parker, 'The Pluralization of Regulation' (2008) 9 *Theoretical Inquiries in Law* 349–368; see also J. Black, 'Decentring Regulation: Understanding the Role of Regulation and Self Regulation in a "Post-Regulatory" World' (1001) 54 *Current Legal Problems* 103, 134–35. Further examples include recommendations by national consumer associations and reports on by national market surveillance authorities, see the *Empfehlungen an Betreiber von Bewertungsplattformen* issued by the German consumer association (*Verbraucherzentrale Bundesverband e.V.*) in December 2012 <www.vzbv.de/sites/default/files/downloads/Bewertungsportale-Empfehlungen-vzbv-2012.pdf> and the Report on online reviews & endorsements published in June 2015 by the UK Competition & Market Authority <https://www.gov.uk/government/uploads/system/uploads/attachment_data/file/436238/Online_reviews_and_endorsements.pdf> accessed 15.06.2016.

authorities. Thus, in 2012 the UK Advertising Standards Authority required the holiday review website TripAdvisor to withdraw misleading claims implying that all the reviews that appeared on the website were from real travellers, or were honest, real or trusted.[34] In a similar case, which also concerned TripAdvisor, the Italian Competition Authority fined the travel review website for publishing misleading information regarding the sources of its online reviews.[35] In a decision of March 2016 concerning a review website for medical practitioners the German *Bundesgerichtshof* (BGH) has clarified the duties of care of the review site operator.[36] The court held that the website operator is obliged to verify the review if a medical practitioners makes a substantiated complaint claiming that the reviewer has not been his patient. According to the BGH the website operator must request documents from the reviewer as evidence for the treatment (e.g. prescriptions, health care vouchers) and forward those documents to the medical practitioner as long as this does not conflict with data protection rules.

3.2. EUROPEAN LEVEL

Recently, reputation systems have caught also the attention of regulators at the European level. In several communications on the Digital Single Market, the European Commission underlines the importance of rating and review mechanisms for the business model of online platforms.[37] The Commission documents also highlight the problem of fake or misleading reviews.[38] So far, however, a certain regulatory reluctance can be observed. In the face of the rapidly changing digital economy the European Commission seems to shy away from suggesting new rules that could stifle innovation.[39] Thus, in its Communication on Online Platforms and the Digital Single Market published in

---

[34] ASA Ruling on TripAdvisor LLC (1 February 2012), Complaint Reference A11-166867 <https://www.asa.org.uk/Rulings/Adjudications/2012/2/TripAdvisor-LLC/SHP_ADJ_166867.aspx#.V2EclWMv_aY> accessed 15.06.2016.

[35] AGCM, 14 December 2014, PS 9345, Provv. No. 25237, TripAdvisor <www.agcm.it/> accessed 15.06.2016; on this case see A. DE FRANCESCHI, 'The Adequacy of Italian Law for the Platform Economy' (2016) *EuCML* 56, 57–58.

[36] BGH, 01.03.2016, Multimedia und Recht (MMR) 2016, 418 – *Ärztebewertung III*; see also BGH, 19.03.2015, MMR 2015, 726 – *Hotelbewertungsportal*; for an overview of recent German case law see D. HÖCH, 'Bewegung bei Bewertungsportalen – wie Unter- nehmen ihren Ruf im Netz besser schützen können' (2016) *Betriebs-Berater* 1475.

[37] See e.g. the European Commission Communication, 'Online Platforms and the Digital Single Market – Opportunities and Challenges for Europe', COM(2016) 288/2 final, p. 11, and European Commission Communication, 'A European agenda for the collaborative economy', COM(2016) 320 final, p. 4.

[38] COM(2016) 288/2 final, p. 11.

[39] On the problem of designing legal rules for disruptive technologies see C. TWIGG-FLESNER (in this volume).

May 2016, the Commission emphasises that the right regulatory framework for the digital economy must be conducive 'to scaling-up of the platform business model in Europe'[40] and 'fostering the innovation-promoting role of platforms'.[41]

Under the new agenda of light-handed regulation (or the 'problem-driven approach'[42] as the Commission prefers to call it), the instruments of choice seem to be self-regulatory or co-regulatory measures. If such instruments are underpinned by appropriate monitoring mechanisms, they can strike – in the view of the Commission – the 'right balance between predictability, flexibility, efficiency, and the need to develop future proof solutions.'[43] This approach has also been expressed in the Commission's recent Communication on the collaborative economy:

> 'When reassessing the justification and proportionality of legislation applicable to the collaborative economy, national authorities should generally take into consideration the specific features of collaborative economy business models and the tools they may put in place to address public policy concerns, for instance in relation to access, quality or safety.'[44]

As one example of such self-regulatory instruments, the Commission explicitly mentions ratings and reputational systems as a tool for reducing risks for consumers stemming from information asymmetries. For example, in its 'European agenda for the collaborative economy' the Commission acknowledges that such reputation mechanisms, 'can contribute to higher quality services and potentially reduce the need for certain elements of regulation, provided adequate trust can be placed in the quality of the reviews and ratings.'[45]

So far, the European Commission has refrained from taking any concrete steps towards enacting specific 'hard law' rules ensuring the integrity of rating and review systems. At the time of writing, there are only two 'soft law' texts that explicitly address this issue. The first of these texts are the 'Key principles for comparison tools'[46] of 2016, which have been elaborated by a multi-stakeholder group launched by the Commission in 2012. These principles have fed into the second text, the recently revised Guidance on the implementation/application

---

[40] COM(2016) 288/2 final, p. 4.
[41] Ibid., p. 5.
[42] Ibid.
[43] Ibid.
[44] European Commission Communication, 'A European agenda for the collaborative economy', COM(2016) 356 final, p. 4.
[45] Ibid., p. 4.
[46] Key principles for comparison tools <http://ec.europa.eu/consumers/consumer_rights/unfair-trade/docs/key_principles_for_comparison_tools_en.pdf> accessed 15.06.2016. These principles have also been referenced in the European Commission's Communication on Online Platforms and the Digital Single Market, COM(2016) 288/2 final.

of the Unfair Commercial Practices Directive (UCPD Guidance),[47] published by the Commission on 25 May 2016.

The UCPD Guidance contains a dedicated chapter on 'user review tools'[48] which gives some indications how the provisions of the UCPD apply to online reputation systems. For example, with regard to the problem of fake reviews posted by a trader (or an e-reputation agency) in the name of consumers the UCPD Guidance makes clear that such practices are contrary to point no. 22 of Annex I of the Directive, which prohibits 'falsely representing oneself as a consumer'.

All in all, the positions taken by the Commission in the UCPD Guidance deserve approval. It is doubtful, however, whether the publication of such a document, which is not legally binding, in combination with the 'Key principles for comparison tool', which have an even more uncertain – not to say obscure – legal nature, is sufficient to provide legal certainty for platform operators and eliminate regulatory grey areas. Moreover, if online rating and review systems are to promote trust in online marketplaces, it is necessary to clearly define the requirements for such reputation mechanisms. Given the crucial role of reputation in the market mechanism, any distortions in the reputation information provided through ratings and reviews may effectively distort the marketplace itself. Therefore, any opacity with regard to the regulatory requirements for reputation systems should be avoided.

4. KEY ELEMENTS OF A REGULATORY FRAMEWORK FOR REPUTATION SYSTEMS

The brief overview of recent regulatory initiatives in the previous section has shown that the regulatory landscape for online consumer review systems at the moment presents a rather uneven picture composed of guidelines, recommendations, standards, legislative proposals and pointillist case law. However, taken together, these different sources reveal some key elements which are necessary for running a functional reputation system. In the terms of social theorist Niklas Luhmann, a well-designed regulatory framework can be a source of 'systemic trust', i.e. confidence in the reliability of the reputational feedback system.[49] Such a systemic trust implies always confidence in the functionality of the built-in control mechanisms.[50] The following section sets out

---

[47] Guidance on the implementation/application of the Unfair Commercial Practices Directive, SWD(2016) 163 final <http://ec.europa.eu/justice/consumer-marketing/files/ucp_guidance_en.pdf> accessed 15.6.2016.
[48] Section 5.2.8 of the UCPD Guidance.
[49] On the notion of 'systemic trust' (*Systemvertrauen*) see N. LUHMANN, *Vertrauen – Ein Mechanismus der Reduktion sozialer Komplexität*, 4th ed., Lucius & Lucius, Stuttgart 2000, pp. 60–79.
[50] *Ibid.*, p. 77.

several elements, which may serve as a basis for building a European regulatory framework for online reputation mechanisms.

4.1. A EUROPEAN STANDARD FOR ONLINE REPUTATION SYSTEMS

Before considering the details of a future regulatory framework, a few words should be said about the appropriate instrument for implementing the quality criteria. Different regulatory techniques can be envisaged. A 'classical' option could be a Directive or a Regulation setting out detailed requirements for reputation systems.[51] However, a more experimental and flexible alternative could be a combination of a Directive, which defines general principles, and a European standard for reputation services elaborated under the auspices of the European Committee for Standardisation (CEN), which spells out the technical details. This regulatory technique has been used for a long time in the area of product safety. Under the so-called 'new approach'[52] to harmonisation introduced in the 1980s directives have been limited to the setting of 'essential requirements'. Technical details are left to standards elaborated by the European standardisation bodies (CEN, CENELEC). The link between the directive and the (voluntary) standards is created via a presumption of conformity. Products that comply with the harmonised standards are presumed to be in conformity with the requirements of the essential health and safety requirements of the corresponding directive.

This method of co-regulation could be transposed to the area of services. This would be very much in the spirit of the new 'Standardisation Package' unveiled on 1 June 2016, in which the European Commission has underlined that standards for services and information and communication technologies (ICT) should be the key priority of the European Standardisation System so that Europe can fully grasp the benefits of the digitalisation and 'servicification' of the economy.[53] A European standard for online reputation systems could be a flexible and innovative regulatory instrument for substantiating both legal and technical aspects of electronic reputation services, combining thus the

---

[51] For an outline of such a 'Platform Directive' addressing also other regulatory issues of the Platform Economy see BUSCH, SCHULTE-NÖLKE, WIEWIÓROWSKA-DOMAGALSKA and ZOLL, *supra* n. 28, pp. 3–10.

[52] European Commission Communication, 'Technical Harmonization and Standardization: A New Approach', COM(1985) 19. The new approach has subsequently been revised; see e.g. European Commission Communication, 'Enhancing the Implementation of the New Approach Directives', COM(2003) 240. For a critical view of see H. SCHEPEL, *The Constitution of Private Governance*, Hart, Oxford 2005.

[53] European Commission Communication, 'European Standards for the 21st century', COM(2016) 358 final, p. 10.

two priorities identified in the Standardisation Package.[54] Such a European standard could take inspiration from the above-mentioned French Standard NF Z 74-501,[55] which currently serves already as a model for the elaboration of a global standard for online reputation systems under the auspices of the International Organisation for Standardization (ISO).[56] Following the model of the 'new approach' a platform operator who implements a reputation system which is in compliance with the European standard would benefit from a presumption of conformity with the regulatory requirements laid down by a European Platform Directive.

Regardless of the preferred regulatory instrument, a European lawmaker should be mindful of the following aspects if he wants to create a set of rules that supports the self-regulatory potential of ratings and review systems and compensates their weaknesses.

4.2. TRANSPARENCY AND CONSUMER INFORMATION

The main function of reputation systems is to improve the informational environment in the online marketplace by providing reputational information about market participants and the goods or services they offer. If a reputation system is to increase market transparency, it is essential that the functioning of the system itself is transparent. Therefore, online platforms should have publicly accessible policies for handling reviews.[57] According to Article 24(3) of the French proposal for the *Loi pour la République numérique* the operator of the reputation system has to provide the platform users with fair, clear and transparent information about the modalities of publication and processing for online reviews. Pursuant to Article 24(4), it must be indicated whether or not the reviews will be subject to any control before publication. This information must also indicate the main characteristics of the control procedure. The Guidelines issued by the Danish Consumer Ombudsman even go further:

---

[54] For a similar approach in the area of Online Dispute Resolution see C. BUSCH and S. REINHOLD, 'Standardisation of Online Dispute Resolution Services: Towards a More Technological Approach' (2015) 4 *EuCML* 50–58; for a general overview see P. DELIMATSIS, *supra* n. 29.

[55] AFNOR, *supra* n. 30.

[56] The International Organization for Standardization (ISO) in 2014 set up a new technical committee (ISO/TC 290 – Online reputation) which is currently elaborating a global standard for methods, tools, processes, measures and best practices related to online reputation of organisations or individuals providing services or products, derived from user-generated content available on the Internet.

[57] See also Section 5 of the Key principles for comparison tools: Comparison tools displaying user reviews should explain that the reviews are user-generated and how they are created, posted, ranked and sorted.

'Terms of use, including basic principles for the procedure for submitting reviews, any use of weighting of reviews and the methods chosen for summarising product reviews, must generally be easily accessible to the general public and to the users of the online service and be structured in an easily comprehensible manner.'[58]

The procedure for submitting of the user reviews should be impartial and ensure that the reviews are processed in an objective manner. It goes without saying that any selection on the basis of the users' positive or negative purchasing experience should not be permitted.[59]

4.3. SUBMISSION OF REVIEWS

Several issues of system design have to be considered with regard to the collection or submission of reviews. The first question is whether the reputation system should be *open or closed*, i.e. whether reviews can be posted by anyone or only after a confirmed transaction. Online platforms take different approaches to this question. Some sharing economy platforms, such as Airbnb, accept reviews only after a confirmed transaction. On other platforms (e.g. Amazon Marketplace) no such restriction applies. Here, everyone who has a registered user account at the platform can write a review for any product. However, reviews based on a confirmed transaction are marked with a small badge ('Verified Purchase'). A third category comprises review platforms that are not linked to an online marketplace (e.g. Yelp, TripAdvisor). On such platforms verification of the user experience is not possible. Therefore, the risk of fake reviews is much higher in these cases.

A regulatory instrument for reputation systems at EU level should leave it to the market to decide whether the system is open or closed. However, as this aspect has a significant influence on the reliability of the reviews, it should be made clear to the platform users who can post a review. This is also the approach taken by the new UCPD Guidance. According to the Commission Staff Working Paper it follows from Articles 6(1)(b) and 7(4)(a) UCPD that 'the platform should not mislead its users as to the origin of the reviews: it should avoid creating the impression that reviews posted through it originate from real users, when it cannot adequately ensure this.'[60] If, however, a platform operator explicitly claims that the reviews originate from users, the UCPD Guidance requires him to take 'reasonable and proportionate steps which … increase the likelihood for such reviews to reflect real users' experiences.' According to the UCPD Guidance, such measures could include requiring reviewers to register, verifying

---

[58] Sec. 6 of the Danish Guidelines on Publication of User Reviews.
[59] Sec. 7 of the Danish Guidelines on Publication of User Reviews.
[60] Sec. 5.2.8 of the UCPD Guidance.

the IP address used to submit the review, or requiring additional evidence (e.g. a booking number) from the reviewer. The duties of the platform operator are limited, however, by Article 15(1) of the E-Commerce Directive according to which Member States shall not impose a general monitoring obligation on providers of information society services.

A matter of controversy is whether *anonymous reviews* should be possible. The admission of anonymous reviews could allow users to be more honest without fearing any direct or indirect retaliation from the other party. As a result, this could improve the quality of the information transmitted through the reputation system. At the same time, anonymous reviews may increase the problem of 'trolling' and the danger of libellous postings. The French standard NF Z 74–501 therefore excludes any anonymous reviews from its scope of application. Similarly, the Guidelines issued by the Danish Consumer Ombudsman require that the reviewer must provide his identity and contact information for use by the operator of the reputation system, so as to ensure that the reviewer can be contacted.[61]

Another issue is the problem of *reciprocal reviewing*. Many sharing economy platforms allow both parties of a transaction (e.g. hosts and renters at Airbnb or drivers and passengers at Uber) to review each other. On the one hand, a system of reciprocal reviewing is a valuable design feature in reputation systems as it builds trust on both sides of the market. On the other hand, it may create strategic incentives for overly positive reviews if reviewers fear retaliation in the event of a negative review.[62] In order to circumvent this problem, Airbnb has recently changed its review system and adopted a system that is sometimes referred to as 'simultaneous reveal': a review is not displayed until the other party has also left a review.[63] In theory, at least, such a simultaneous-reveal system will reduce problems associated with strategic considerations in reviewing. If empirical research confirms this assumption, one might consider requiring 'simultaneous reveal' as a standard for systems that allow parties to review each other.

This raises the question whether such a design feature should be left to the market. The answer can be positive, if there are sufficient incentives for the platform operator to choose a system design that increases the reliability of reviews. On the one hand, operators of platform markets have an interest in maintaining the integrity of their reputation system. A totally dysfunctional reputation system would eventually translate into a market failure and thus endanger the business model of the platform. On the other hand, there is

---

[61] Sec. 6 of the Danish Guidelines on Publication of User Reviews. For clarification, the Guidelines explicitly state that it would not be contrary to the Danish Marketing Practices Act if the terms of use of a reputation system allow reviewers to remain anonymous in *published* reviews.

[62] G. BOLTON, B. GREINER and A. OCKENFELS, 'Engineering Trust: Reciprocity in the Production of Reputation Information' (2013) 59 *Management Science* 265–285.

[63] M. LUCA, *supra* n. 15.

an incentive for platform operators to paint the picture just a little bit more rosy than the reality and to accept a certain degree of 'rating inflation' on the platform. Therefore, a regulatory intervention imposing a 'simultaneous reveal' standard seems to be justified.

Another controversial issue is the admissibility of *paid reviews*, i.e. reviews that are posted in exchange for compensation of any kind. Economic studies of the 'market for evaluations' suggest that reviews of goods and services constitute public goods.[64] Each user of an online marketplace can benefit from an evaluation without reducing its value for other users. As a consequence, there is a tendency that voluntary reviews are underprovided, leading to a suboptimal supply of evaluations. A small incentive (e.g. entry into a contest or sweepstakes, discounts on future purchases) for posting a review could increase the number of reviews and thus increase the information available in the marketplace. However, such incentives could result in biased reviews and thus mislead other users. There seems to be broad consensus that buying positive reviews is not admissible.[65] In contrast, neither the French standard NF Z 74-501 nor the Guidelines issued by the Danish Consumer Ombudsman provide for a blanket ban of reviews against payment or other benefit. However, both formulate transparency requirements for such reviews.[66] For example, the comments to Section 7 of the Danish Guidelines suggest adding an explicit reference indicating that the reviewers have been entered into a competition in return for submitting their review.

4.4. PROCESSING OF REVIEWS

After submission, the reviews usually undergo some sort of processing in order to ensure that the reviews are in accordance with the general terms of use of the platform. This raises the question under which conditions the platform operator may be liable for overtly inappropriate comments, harassment or libellous reviews.

From an EU law perspective, the platform operator may benefit from the liability exemption for hosting providers under Article 14(1) of the E-Commerce Directive. According to the case law of the CJEU this provision 'must be interpreted as applying to the operator of an online marketplace where that operator has not played an *active role* allowing it to have knowledge or control of

---

[64] C. AVERY, P. RESNICK and R. ZECKHAUSER, 'The Market for Evaluations' (1999) 89 *American Economic Review* 564, 565.
[65] See Sec. 4.3.3 of the French standard NF Z 74-501 and the comments to Sec. 7 of the Danish Guidelines on Publication of User Reviews.
[66] See Sec. 4.3.4 of the French standard NF Z 74-501 and Sec. 7 of the Danish Guidelines on Publication of User Reviews; see also Sec. 5 of the Key principles for comparison tools.

the data stored.'⁶⁷ Moreover, Article 15(1) of the E-Commerce Directive prevents Member States from imposing on such 'hosting service providers' a general *obligation to monitor* the stored information or to actively engage in fact-finding. At the same time, recital 48 of the E-Commerce Directive underlines that the Directive 'does not affect the possibility for Member States of requiring service providers, who host information provided by recipients of their service, to apply *duties of care*, which can reasonably be expected from them and which are specified by national law, in order to detect and prevent certain types of illegal activities.'

In applying these principles, the German BGH in a ruling of 19 March 2015 held that the operator of HolidayCheck, a hotel review site, was not responsible for a review posted by a user stating that 'for 37.50 EUR per person per night there were bedbugs' in a specific hotel.⁶⁸ The BGH dismissed the hotel owner's claim for damages. The court held that the operator of the review site had not taken an active role, but rather a passive or *neutral role* in relation to the reviews. According to the BGH the operator of the review system did not leave this neutral role when he applied an automated statistical processing and a word filter to the reviews in order to ensure conformity with the platform's terms of use. The court concluded that the operator of the review site would only violate his duty of care and thus be liable if he has been notified of any illegal activity and failed to act expeditiously to remove the illegal content and to take measures to prevent similar incidents in the future.

Although a platform operator has no general duty to monitor the reviews, it is common practice that reviews undergo a fully or partly automated processing in order to ensure that the reviews are in accordance with the general terms of use of the platform. Automated processing is also used for example for deleting fake reviews, sometimes referred to as 'opinion spamming', which can be identified through algorithms that mine review texts for specific hints (e.g. duplicate or near-duplicate reviews, use of abnormal text patterns).⁶⁹ Both the Danish Guidelines and the French standard NF Z 74–501 contain a non-exhaustive list of reasons for rejecting a user review.⁷⁰ Examples include incomprehensible reviews and overtly inappropriate comments, but also reviews containing personal information which may potentially be used for identity theft. If a review is rejected, the reviewer must be informed about the rejection and the reasons for such rejection. This is a common principle of

---

67   CJEU, 12.07.2011, Case C-324/09 (*L'Oréal SA*), para. 123.
68   BGH, 19.03.2015, MMR 2015, 726 – *Hotelbewertungsportal*.
69   See e.g. N. JINDAL and B. LIU, 'Opinion spam and analysis', Proceedings of the 2008 International Conference on Web Search and Data Mining, ACM, New York, 2008, pp. 219–230; see also J. HIRSCHBERG and C.D. MANNING, 'Advances in natural language processing' (2015) 349 *Science* 261–266.
70   Sec. 8 of the Danish Guidelines on Publication of User Reviews; see also Sec. 5.3 of the French standard NF Z 74–501.

procedural fairness which has been acknowledged by the Danish Guidelines,[71] the French Standard[72] and also Article 24(6) of the French proposal for the *Loi pour la République numérique*.

4.5. PUBLICATION OF REVIEWS

Another point to be considered is the presentation of the reviews on the platform. According to the Guidelines issued by the Danish Consumer Ombudsman, all reviews, whether positive or negative, must be presented in chronological order or in another objective manner.[73] Similarly, the French standard requires a chronological display of the reviews as a default setting.[74]

A good reputation system should quickly respond to changes in the quality level offered by a seller or service provider on the platform. Therefore it is necessary that reviews are published without undue delay.[75] As a matter of transparency, the time limit for publication of a review should be indicated in the terms and conditions of the platform. The French standard NF Z 74–501 goes even further and stipulates that reviews shall be published within one month of their submission date.[76] Again it goes without saying that the time limit shall be the same for positive and negative reviews.[77]

In order to assess whether a review is still relevant, it is necessary to know when the reviewer experienced the goods or services. The older the review, the less likely it is that it still represents the current quality of the goods or service. Therefore, Article 24(5) of the French proposal for the *Loi pour la République numérique* requires that the date of review is indicated (and, as the case may be, the date of any updates). Similarly, the introduction of an 'expiry date' for reviews may be useful. An example of such a rule can be found in the Guidelines issued by the Danish Consumer Ombudsman. Section 9 of the Guidelines provides that the platform operator may determine a time limit for removal of published reviews. This time limit may vary depending on the product category. The minimum time limit defined by the Danish Guidelines is 12 months. The existence of an 'expiry date' for reviews is not only relevant for ensuring that the reputational information is up to date – it may also reduce entry barriers for new market participants.

---

[71] Sec. 8 of the Danish Guidelines on Publication of User Reviews.
[72] Sec. 5.2.3 of the French standard NF Z 74–501.
[73] Sec. 9 of the Danish Guidelines on Publication of User Reviews.
[74] Sec. 6.1. b) of the French standard NF Z 74–501.
[75] Sec. 9 of the Danish Guidelines on Publication of User Reviews; see also Sec. 6.1. d) of the French standard NF Z 74–501.
[76] Sec. 6.2.4 of the French standard NF Z 74–501.
[77] Sec. 9 of the Danish Guidelines on Publication of User Reviews; see also Sec. 6.1. d) of the French standard NF Z 74–501.

4.6. CONSOLIDATED RATINGS

On many platforms individual user reviews are consolidated into overall ratings, often expressed in 'stars' or scores. These consolidated ratings are particularly important in influencing consumers' purchasing decisions.[78] Considering the limits of human attention, such consolidated ratings can help to mitigate the problem of information overload which could be caused by a large number of confusing and contradictory reviews. The use of consolidated ratings thus takes into consideration the problems of bounded attention and bounded rationality and increases the salience (i.e. the cognitive accessibility) of the most important information. In this sense, consolidated ratings are similar to other forms of behaviourally informed disclosures using a one-dimensional summary of one or more product features in the form of a 'score'. A well-known example is the annual percentage rate (APR), which provides aggregated information about the total cost of a consumer credit, including any related costs and fees.[79]

However, problems of bounded attention and bounded rationality also occur with regard to consolidated ratings. Empirical research shows that many platform users pay attention to the average rating, but not to the *number* of ratings or to the combination of the average and the number of ratings together.[80] As a consequence, it is not unlikely that consumer behaviour is influenced by an average rating which is based on a rather small number of user experiences, i.e. an insufficient sample size. Both the French standard NF Z 74–501 and the Guidelines issued by the Danish Consumer Ombudsman address this issue and require that in cases where user reviews are consolidated in an overall rating, the operator of the reputation system must provide general and clear information about how the rating has been determined and the total number of reviews on which the rating is based.[81]

The simplicity of overall ratings or scores improves not only transparency but also comparability of products. At the same time, the one-dimensional nature of scores or consolidated ratings is one of the weaknesses of this model.[82] As scores are based on a certain model of information aggregation and necessarily

---

[78] See e.g. B. DE LANGUE, P.M. FERNBACH and D.R. LICHTENSTEIN, 'Navigating by the Stars: Investigating the Actual and Perceived Validity of Online User Ratings' (2016) 42 *Journal of Consumer Research* 817.

[79] Article 4(2)(c) of the Consumer Credit Directive (2008/48/EC) and Article 11(2)(3) of the Mortgage Credit Directive (2014/17/EU).

[80] A.J. FLANAGIN, M.J. METZGER, R. PURE, A. MARKOV and E. HARTSELL, 'Mitigating risk in ecommerce transactions: perceptions of information credibility and the role of user-generated ratings in product quality and purchase intention' (2014) 14 *Electronic Commerce Research* 1.

[81] See Sec. 9 of the Danish Guidelines on Publication of User Reviews and Sec. 6.2.2 of the French standard NF Z 74–501.

[82] O. BAR-GILL, 'Defending (Smart) Disclosure: A Comment on More Than You Wanted to Know' (2015) 11 *Jerusalem Review of Legal Studies* 75. .

imply some sort of selective disclosure, their relevance depends on the algorithm used to calculate the consolidated rating. For example, it makes a difference whether the calculation of the overall rating is based only on a selection of recent reviews or whether all reviews are taken into account without any time limit. While the first approach may provide a more up-to-date impression, the latter approach makes sure that the average rating is calculated on a broader statistical basis.[83] Similarly, it could be envisaged that the rating algorithm gives different weight to different reviews. In this sense, reviews by verified customers, reviews from reviewers with a longer transaction history or reviews that have been flagged as 'helpful' by other platform users might be given a greater weighting in the calculation of the overall score. On the one hand, such an approach could increase the informational value and thus the relevance of the overall rating. On the other hand, such a black-box model for reputation scoring is also vulnerable to manipulation or discrimination by the platform operator.[84] For example, it would be problematic if the aggregator were to take into account geolocation data of the reviewer in order to give different weight to the opinions expressed by reviewers from different Member States.

This raises the question whether the operator of a reputation system should be obliged to disclose their rating algorithms if the impartiality of a consolidated rating is questioned. In a recent decision concerning the review platform Yelp, the Higher Regional Court of Berlin answered this question in the negative.[85] The court held that the operator of the review platform is under no obligation to disclose the details of its recommender software as such details are protected as business secrets. Moreover, the court underlined that in order to effectively combat manipulation of the review system it is necessary that the users do not know how exactly a review is evaluated. The decision of the Berlin court is in line with an earlier decision of the *Bundesgerichtshof* in which the court held that a credit agency is not obliged to disclose its 'score formula' and the methodology for calculating credit scores to an individual requesting access to such information on the basis of the German Data Protection Act.[86]

On the one hand, both decisions are understandable from a technical perspective. Not only would a disclosure of the rating algorithm raise

---

[83] According to Sec. 6.2.2. of the French standard NF Z 74-501 the overall rating for a product or service shall only take into account ratings in reviews submitted within a time period shorter than or equivalent to the 'expiry limit' defined by the operator of the reputation system. For the hotel and restaurant sector, the standard fixes the expiry limit to two years.

[84] See F. PASQUALE, 'Reputation Regulation: Disclosure and the Challenge of Clandestinely Commensurating Computing' in S. LEVMORE and M.C. NUSSBAUM (eds.), *The Offensive Internet: Privacy, Speech, and Reputation*, Harvard University Press, Cambridge MA 2010, p. 107, 117–120; see also F. PASQUALE, *The Black Box Society: The Secret Algorithms that Control Money and Information*, Harvard University Press, Cambridge MA 2015.

[85] Kammergericht, 10.12.2015, *Multimedia & Recht* 2016, 353.

[86] BGH, 28.01.2014, NJW 2014, 1235; for a critical view of the decision see T. WEICHERT, 'Scoring in Zeiten von Big Data' [2014] *Zeitschrift für Rechtspolitik* 168–171.

commercial sensitivities and issues of intellectual property, it might also weaken the reputation system itself as it is easier to game a system if the architecture of the system is transparent. On the other hand, the lack of transparency of the rating algorithm leaves a major gap in the system of legal protection of platform users.

4.7. RIGHT OF REPLY

A general feature of the fairness requirements for reputation system is a 'right of reply', allowing users whose goods or services have been the object of a review to tell their side of the story.[87] According to Article 24(7) of the French proposal for the *Loi pour la République numérique*, the operator of the reputation system must provide a free-of-charge complaint mechanism which allows a user whose goods or services have been the object of a review to submit a reasoned notification if he has doubts regarding the authenticity of the review. A similar rule is contained in Section 9 of the Guidelines issued by the Danish Consumer Ombudsman.

While the French legislative proposal does not spell out further details of the complaint procedure, the Danish Guidelines require that objections must be processed as soon as possible in accordance with the platform operator's terms of use. Moreover, during the processing period, the platform operator may opt to mark or suspend the objections and the reviews involved. The French standard NF Z 74–501 defines even more rigid requirements and stipulates that the replies must be published within a maximum of seven calendar days from the request, and shall be linked to the review in question.[88]

5. CONCLUSION

Reputation systems play a key role in building trust among strangers using collaborative economy platforms. Almost all sharing economy platforms use these reputational mechanisms, such as ratings and reviews, to establish trust in the marketplace. From a regulatory perspective, the growing importance of online reputation systems gives rise to a dual challenge. On the one hand, a well-functioning reputation system can complement more traditional forms of consumer protection such as occupational licensing and mandatory disclosures. On the other hand, deficiencies in the system design of reputation mechanisms

---

[87] See also E. GOLDMAN, 'Regulating Reputation' in H. MASUM and M. TOVEY, *The Reputation Society*, MIT Press, Cambridge MA 2011, p. 51, 56–57.
[88] Sec. 6.1. d) of the French standard NF Z 74–501.

could mislead consumers and cause severe market inefficiencies in the collaborative economy.

Therefore, it is necessary to set up a regulatory framework that ensures the effectiveness of trust building through reputation systems on the basis of clear quality criteria. Recent regulatory initiatives at Member State level could further increase regulatory fragmentation in in the area of online platforms and create new barriers in the European single market. In order to promote the development of the collaborative economy and to ensure a level playing field for online platforms, it is necessary to create harmonised European rules for online rating and review system. Following the model of the 'new approach' which has been efficient and successful in the area of product safety, a flexible and innovative regulator strategy could be a combination of a Directive, which defines 'essential requirements' for trustworthy and fair reputational feedback systems, and a harmonised European service standard which spells out the legal and technical details.

ONLINE DISPUTE RESOLUTION PLATFORM

Making European Contract Law More Effective*

Jorge Morais Carvalho and Joana Campos Carvalho

1. Introduction 245
2. Overview of Alternative Dispute Resolution in European Contract Law 247
3. Online Dispute Resolution Platform............................ 250
    3.1. Disputes Covered by Regulation 524/2013 250
    3.2. ADR Entities 251
    3.3. Consumer ODR Procedures 253
    3.4. Costs of Use of the Platform and of Consumer ADR Procedures ... 256
    3.5. Common Rules Applicable to the ADR Procedure within the ODR Platform... 257
4. Critical Analysis of the Legal Regime and its Implementation........ 263
    4.1. Consumer Information 263
    4.2. Mandatory ADR... 265
    4.3. Persuasive Effect of the ADR Entity's Intervention.............. 266

1. INTRODUCTION

Contract law and civil procedure law are two sides of the same coin. The absence of adequate means to apply the national and European substantive rules renders these rules irrelevant.[1]

One of the main goals of the EU legislation related to contract law is ensuring the functioning of the internal market, which shall comprise an area without

---

* We would like to thank Prof. Alberto De Franceschi for the invitation to participate in the European Contract Law Seminars as well as for the warm welcome in Ferrara.
[1] J. Luzak, 'The ADR Directive: Designed to Fail? A Hole-Ridden Stairway to Consumer Justice' (2016) 24 *ERPL (European Review of Private Law)*, 81, 81.

internal frontiers in which goods, persons, services and capital move freely (Article 26(2) Treaty on the Functioning of the European Union (TFEU)). In order to achieve this, the European Union shall adopt, *inter alia*, laws which ensure a high level of consumer protection (Articles 114 and 169 TFEU).

However, a high-level legal framework for consumer contracts may not be enough as the enforceability of a legal framework depends on procedures that allow the parties, especially the consumer, to settle disputes arising throughout the life of a contract.

At a European level, the existence of simple, efficient, fast and low-cost ways of resolving disputes is key to making European contract law – which is mainly consumer law – more effective. This idea inspired the legislative package adopted by the EU in 21 May 2013, which includes Directive 2013/11/EU of the European Parliament and of the Council on alternative dispute resolution for consumer disputes (Directive on Consumer ADR) and Regulation (EU) 524/2013 of the European Parliament and of the Council on online dispute resolution for consumer disputes (Regulation on Consumer ODR).

For example, if after calling an Uber, the car does not show up and the price of €3 is charged, would anyone even consider going to court?

The references to dispute resolution are nevertheless scarce in the supporting documents of the European digital single market strategy.

The European Commission, in a Communication of 6 May 2015, named 'A Digital Single Market Strategy for Europe',[2] confirms our idea, stating that 'just having a common set of rules is not enough' and emphasising the 'need for more rapid, agile and consistent enforcement of consumer rules for online and digital purchases to make them fully effective'. It is also added that 'the Commission will establish in 2016 a EU-wide online dispute resolution platform'.

In another Communication, of 9 December 2015, named 'Digital contracts for Europe – Unleashing the potential of e-commerce',[3] the Commission refers that the entry into operation of the Online Dispute Resolution platform will, among other measures, 'enhance cross-border commerce and remedy *other important problems* for business and consumers, such as the need for effective cross-border redress and enforcement'.

However, enforcement of substantive rules is not only *another important problem* for consumers. It is a key problem. We admit that the problem may not be identical in all Member States,[4] in view of their diversity and of the diversity of the rules on access to courts and justice in general, but it is possible to state

---

[2] COM(2015) 192 final.
[3] COM(2015) 633 final.
[4] M.B.M. Loos, 'Enforcing Consumer Rights through ADR at the Detriment of Consumer Law' (2016) 24 *ERPL* 61 has a very critical view of the Directive. In many Member States the statement that 'in practice the ADR procedure need not be much more shorter than an ordinary court procedure would have been' (p. 66) will not be accurate. First the vast majority of consumer cases will never be presented before an ordinary court. Second even with a few

with strong conviction that there are cross-cutting issues in what regards an effective enforcement of European contract law. This is especially relevant in consumer disputes, taking into consideration that they are in most cases of low value when looked at in isolation.

Rules on enforcement are very scarce in the digital single market strategy package. Article 18(1) of the Proposal for a Directive on certain aspects concerning contracts for the supply of digital content[5] states that 'Member States shall ensure that adequate and effective means exist to ensure compliance with this Directive'. The same rule can be found in Article 17(1) of the Proposal for a Directive on certain aspects concerning contracts for the online and other distance sales of goods.[6] Perhaps the idea was that the legislation on consumer ADR already provides on the issue, but these two rules seem clearly not enough in a wider perspective of a digital single market strategy.

In another perspective, in a world where almost everything is online, dispute resolution still resists this tendency.[7] The ODR platform can be a useful tool to help change this paradigm.

This article focuses precisely on the ODR platform created by Regulation 524/2013, which is in operation since 15 February 2016 in most Member States. The article starts with an overview of alternative dispute resolution in European contract law. Secondly, we carry out a legal analysis of Regulation 524/2013 emphasising the main legal issues arising from the ODR platform. Finally, we carry out a critical evaluation of the legal regime and its practical implementation, raising issues and pointing out possible solutions.

2. OVERVIEW OF ALTERNATIVE DISPUTE RESOLUTION IN EUROPEAN CONTRACT LAW

Besides the traditional means, namely the judicial courts, disputes can also be solved through alternative dispute resolution (ADR) means, like mediation and arbitration. The specificities of consumer disputes encourage the use of ADR means.

We can identify three stages in the evolution of the ADR in the European Union:[8]

---

extensions the ADR procedure will certainly be in average much shorter than an ordinary court procedure.
5   COM(2015) 634 final.
6   COM(2015) 635 final.
7   M.A. BULINSKI AND J.J. PRESCOTT, 'Online Case Resolution Systems: Enhancing Access, Fairness, Accuracy, and Efficiency' (2016) 21 *Michigan Journal of Race & Law* 205, 208.
8   On this topic, see I. BENÖHR, 'Alternative Dispute Resolution for Consumers in the European Union' in C. HODGES, I. BENÖHR and N. CREUTZFELDT-BANDA (eds.), *Consumer ADR in Europe*, Hart – C.H. Beck – Nomos, Oxford 2012, p. 1, 7–13.

- first stage: problem identification and approval of non-binding instruments;
- second stage: approval of binding sector-specific instruments; and
- third stage: approval of binding consumer ADR instruments.

The first stage began with Commission Green Paper of 16 November 1993 on access of consumers to justice and the settlement of consumer disputes in the single market.[9] In this document, the Commission made reference to consumer ADR means, emphasising their existence in most of the Member States.

A few years later Commission Recommendation 98/257/EC of 30 March 1998 on the principles applicable to the bodies responsible for out-of-court settlement of consumer disputes is adopted. It applies only to bodies which propose or impose a solution. It was published together with a Communication from the Commission on the out-of-court settlement of consumer disputes.[10]

The Recommendation includes six principles that need to be respected by the persons or bodies who promote consumer ADR: independence; transparency; adversarial principle; effectiveness; legality; liberty and representation.

Commission Recommendation of 4 April 2001 on the principles for out-of-court bodies involved in the consensual resolution of consumer disputes is mainly applicable to all consumer ADR means, irrespective of the nature of the intervention of the out-of-court body. The Recommendation sets out four minimal guarantees that the bodies responsible for the out-of-court settlement must respect: impartiality; transparency; effectiveness and fairness.

The second stage was marked by the approval of binding instruments.

In some directives, Member States are encouraged to create ADR means. The Directive on electronic commerce (Directive 2000/31/EC) has a rule on out-of-court dispute settlement (Article 17). The same happened in other directives, like the Directive on timeshare (Directive 2008/122/EC), whose Article 14(2) states that 'Member States shall encourage the setting up or development of adequate and effective out-of-court complaints and redress procedures for the settlement of consumer disputes under this Directive and shall, where appropriate, encourage traders and their branch organisations to inform consumers of the availability of such procedures'.

In other cases, EU law goes further, imposing adequate and effective ADR means. See Directives on payment services (Directive 2007/64/EC) and on credit agreements for consumers (Directive 2008/48/EC) or Directives in the energy sector (2009/72/EC) and in the telecommunications sector (2009/136/EC). For example, Article 24 of Directive 2008/48/EC states that 'Member States shall ensure that adequate and effective out-of-court dispute resolution procedures for the settlement of consumer disputes concerning credit agreements are put in place, using existing bodies where appropriate' and that 'Member States

---

[9] COM(93) 576 final.
[10] COM(1998) 198 final.

shall encourage those bodies to cooperate in order to also resolve cross-border disputes concerning credit agreements'.

This second stage is also marked by Directive 2008/52/EC of the European Parliament and of the Council of 21 May 2008 on certain aspects of mediation in civil and commercial matters. It is not an economic-sector-specific directive but a legal instrument which specifically regulates one of the ADR means, harmonising some rules at a European level.

Throughout this second stage, the Court of Justice of the European Union (CJEU) has been called upon to give his opinion in a few cases on ADR.[11]

The Italian system of mandatory mediation was deemed compatible with European law (Joined Cases C-317/08 to C-320/08).

The CJEU also concluded that consumers are not obliged to using ADR means. Contractual terms which impose ADR have been deemed unfair (C-168/05; C-40/08).

The third stage began with the Communication from the Commission of 13 April 2011, named 'Twelve levers to boost growth and strengthen confidence'. One of the twelve levers was the approval of legislation on ADR.

Following this Communication, the Commission presented the Proposal for a Directive on alternative dispute resolution for consumer disputes[12] and the Proposal for a Regulation on online dispute resolution for consumer disputes.[13]

These proposals led to the approval of Directive 2013/11/EU of 21 May 2013 on alternative dispute resolution for consumer disputes (Directive on Consumer ADR) and Regulation (EU) 524/2013 of 21 May 2013 on online dispute resolution for consumer disputes (Regulation on Consumer ODR).

In this chapter we will focus mainly on the Regulation, referring to the Directive only incidentally. In relation to the Directive specifically, however, it is important to highlight the four key elements of the European rules on consumer ADR.

Firstly, the main EU goal is to ensure the existence of ADR means for all consumer disputes in all the Member States. Recital 7 of the Directive is clear when stating that 'in order for consumers to exploit fully the potential of the internal market, ADR should be available for all types of domestic and cross-border disputes covered by this Directive'.[14]

Secondly, traders must inform consumers of the consumer ADR means available. Article 13(1) of the Directive states that 'Member States shall ensure that traders established on their territories inform consumers about the ADR entity or ADR entities by which those traders are covered, when those traders

---

[11] H.-W. MICKLITZ, 'European Private Law' in D. PATTERSON and A. SÖDERSTEN (eds.), *A Companion to European Union Law and International Law*, Wiley Blackwell, 2016, p. 262, 275.
[12] COM(2011) 793.
[13] COM(2011) 794.
[14] On this issue, especially with connection with the ODR platform, see section 3.2.

commit to or are obliged to use those entities to resolve disputes with consumers'. Regulation 524/2013 has a similar rule, which will be analysed in section 4.1.

Thirdly, the quality of ADR entities is guaranteed by the respect of six fundamental principles (expertise; independence; impartiality; transparency; fairness and effectiveness), reflected in a set of clear and objective rules for easier verification of compliance.

Finally, the Directive provides for an intensive monitoring of the activity of consumer ADR entities by Member States in order to verify that they are complying with the quality rules referred to in the preceding paragraph.

3. ONLINE DISPUTE RESOLUTION PLATFORM

The ODR platform, created and run by the European Commission as provided for by Regulation 524/2013, has been in operation since 15 February 2016,[15] at the following address: http://ec.europa.eu/consumers/odr/.

Regulation 524/2013 is complemented by Commission Implementing Regulation (EU) 2015/1051 of 1 July 2015 on the modalities for the exercise of the functions of the online dispute resolution platform, on the modalities of the electronic complaint form and on the modalities of the cooperation between contact points.

3.1. DISPUTES COVERED BY REGULATION 524/2013

Article 2(1) of the Regulation sets forth that it shall apply to disputes concerning contractual obligations stemming from online sales or service contracts, regardless of whether they are national or cross-border. This means, on the one hand, that a dispute between two parties from different Member States arising from a contract concluded in person is not covered by the Regulation and, on the other hand, that a contract concluded online by two parties of the same Member State is included in its scope of application.

It should be noted that if one of the parties is resident or established outside the European Union the platform cannot be used.

Article 2(1) also determines that the platform can be used by the consumers in disputes initiated against a trader. The definition of 'consumer' that is relevant for this purpose is the definition in Article 4(1)(a) Directive 2013/11/EU: 'any natural person who is acting for purposes which are outside his trade, business, craft or profession'. Recital 18 of the Directive clarifies that, 'if the contract is

---

[15] According to the information available at the platform's homepage, on 16.05.2016, there were no ADR entities available for certain sectors and in the following countries: Croatia, Spain, Luxembourg, Poland and Romania.

concluded for purposes partly within and partly outside the person's trade (dual purpose contracts) and the trade purpose is so limited as not to be predominant in the overall context of the supply, that person should also be considered as a consumer'.[16]

According to Article 2(2), the Regulation also applies to disputes initiated by a trader against a consumer. However, this possibility is allowed only in 'so far as the legislation of the Member State where the consumer is habitually resident allows for such disputes to be resolved through the intervention of an ADR entity'. The Member States shall provide that information to the European Commission and, where national legislation allows for the resolution of such disputes through ADR, inform the Commission about which ADR entities deal with those disputes. Even if the Regulation applies to disputes initiated by a trader against a consumer, the Member State is not obliged to ensure that ADR entities offer procedures for the out-of-court resolution of such disputes (Art. 2(4)).

In practice, ADR entities included in the platform only deal with disputes initiated by consumers against traders.

Opening up the possibility to traders of using ADR entities to solve disputes against consumers could, bearing in mind the low cost of such procedures, turn ADR entities into debt collection entities. Typically, claims from traders against consumers merely concern the collection of the price of the good or service provided, which means there is not really a dispute concerning the facts or the law applicable to the case. Using ADR entities for that purpose, namely to obtain a binding decision that is enforceable, in a short period of time and at a low cost, could block up the whole system (because of the large number of cases) and would undermine the real purpose of ADR, which is to allow for the resolution of disputes where there is a real conflict of interests between the parties. This does not mean there cannot be alternative dispute resolution procedures in cases where there is no conflict, namely in cases of over-indebtedness, resulting from the accumulation of debt by consumers, but these procedures are usually initiated by the consumer who is in difficulty and not by the trader.

3.2. ADR ENTITIES

The Regulation does not include an autonomous definition of ADR entity, referring instead to the definition of Directive 2013/11/EU.

Article 4(1)(h) of the Directive defines an ADR entity as 'any entity, however named or referred to, which is established on a durable basis and offers the resolution of a dispute through an ADR procedure and that is listed

---

[16] J. Morais Carvalho, *Manual de Direito do Consumo*, 3rd ed., Almedina, Coimbra 2016, pp. 17–23.

in accordance with article 20(2)'. The list to which this rule refers shall be compiled by the competent national authorities and shall include, among others: the name, contact details and website addresses of the ADR entity; the fees it charges, if applicable; the language or languages in which complaints can be submitted and the ADR procedure conducted; the types of disputes covered by the ADR procedure; the sectors and categories of disputes covered by each ADR entity; whether the parties or their representatives (if applicable) need to be physically present; the binding or non-binding nature of the outcome of the procedure; and the grounds on which the ADR entity may refuse to deal with a given dispute.

This information is used by the ODR platform, given that Article 2(1) *in fine* of the Regulation refers to these lists (one for each Member State). The entities included in those lists by the national competent authorities of the Member States must also necessarily be included in the ODR platform. The lists therefore have a double purpose: on the one hand, they provide information on which ADR entities comply with the standards set forth by the Directive and are therefore able to provide ADR procedures, whether internal or cross-border, regardless of whether or not the disputes concern contracts concluded through the internet; and on the other hand, they define which entities take part in the ODR platform.

The platform provides the public with information on the ADR entities included in the list that are competent to handle the disputes included in the scope of application of the Regulation and therefore also included in the scope of application of the platform on which these entities are electronically registered.

Recital 23 of the Regulation highlights that ensuring that all ADR entities listed 'are registered with the ODR platform should allow for full coverage in online out-of-court resolution for disputes arising from online sales or service contracts'. In fact, the Directive aims for the full coverage of ODR procedures in all Member States. However, as we have already pointed out, at least in relation to the ODR platform, as at May 2016 there were still no ADR entities available for some sectors and in some countries.

The list is alphabetically sorted and may be found at: https://webgate.ec.europa.eu/odr/main/index.cfm?event=main.adr.show. It is however, also possible to conduct a search by country. Besides naming the entities, the list also includes a brief description of each one, making reference to the elements that must be included in the lists compiled by the Member States, as mentioned above.

An analysis of the list shows a wide diversity among the different Member States. This diversity is shown by the number of ADR entities available in the Member States, ranging from 38 in the United Kingdom and 18 in Denmark to only one in Cyprus, Ireland and Slovenia. This difference does not allow for any conclusions to be drawn regarding the quality of the ADR entity/ies in those countries, but only reflects the differences concerning the structural organisation of the ADR system in each country.

Figure 1. ADR entities on the ODR platform by Member State

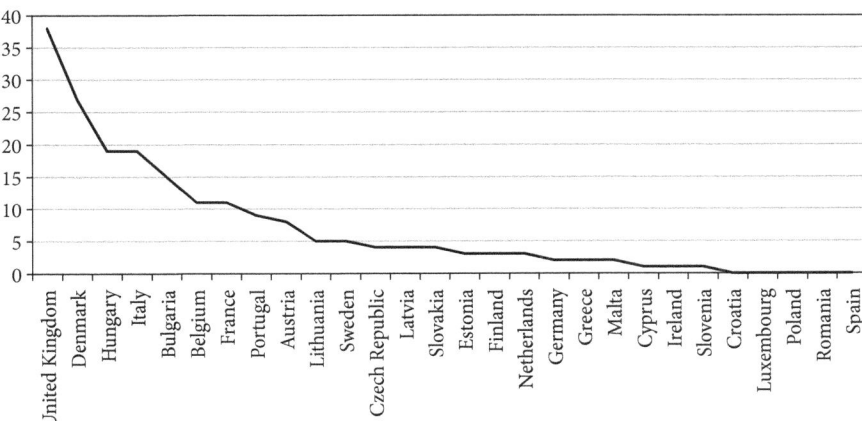

The diversity among Member States is also shown by the sectorial or non-sectorial nature of the ADR entities. For instance, in Austria or in Denmark almost all entities have sectorial competence, while in Portugal the tendency is the opposite (only two out of nine entities are sectorial; the other entities have competence to deal with all consumer disputes, although limited by their territorial competence).

While in some countries those entities are created exclusively to handle the resolution of consumer disputes (e.g. in Portugal), in others we can find many entities that do not have the resolution of consumer disputes as a sole purpose (in Italy several entities named on the list are chambers of commerce).

The dispute resolution procedures also vary significantly from country to country. While in the UK, for instance, ombudsman services are preferred and in Belgium we find a combination of ombudsman services and entities that provide mediation, thus with non-binding results, in Portugal all entities named in the list offer both mediation services (non-binding) and arbitration services (binding).

3.3. CONSUMER ODR PROCEDURES

Relating to the previous section, it should be noted that, strangely, the type of procedure offered is not one of the pieces of information that must be included for each of the entities. This omission is perhaps explained by the difficulties of qualification from a European point of view, given the variety of procedures and terminologies.[17]

Supporting this view, recital 21 of the Directive states that 'ADR procedures are highly diverse across the Union and within Member States. They can take the form of procedures where the ADR entity brings the parties together with the aim of facilitating an amicable solution, or procedures where the ADR entity proposes a solution or procedures where the ADR entity imposes a solution. They can also take the form of a combination of two or more such procedures. This Directive should be without prejudice to the form which ADR procedures take in the Member States'. Recital 22 goes a bit further, admitting, if Member States so determine, procedures 'before dispute resolution entities where the natural persons in charge of dispute resolution are employed or receive any form of remuneration exclusively from the trader', as long as those entities 'are in complete conformity with the specific requirements on independence and impartiality laid down in this Directive'. Recital 23 points in the direction of the exclusion of 'procedures before consumer-complaint handling systems operated by the trader', of 'direct negotiations between the parties' and of 'attempts made by a judge to settle a dispute in the course of a judicial proceeding concerning that dispute'.

Having laid down what consumer ADR procedures are not or cannot be, it is important to also mention that both the Directive – see recital 19 – and the Regulation – see Article 3 – expressly state that they should be without prejudice to Directive 2008/52/EC of the European Parliament and of the Council of 21 May 2008 on certain aspects of mediation in civil and commercial matters. This clearly points towards the recognition of mediation as a consumer ADR procedure,[18] despite the fact that the word mediation is not mentioned once in the Regulation and is mentioned only three times in the Directive, in recital 19. Indeed, we find several entities registered in the ODR platform that include the word 'mediation' in their name.[19]

Directive 2008/52/EC defines mediation as 'a structured process, however named or referred to, whereby two or more parties to a dispute attempt by themselves, on a voluntary basis, to reach an agreement on the settlement of their dispute with the assistance of a mediator.[20] This process may be initiated

---

[17] C. HODGES, I. BENÖHR and N. CREUTZFELDT-BANDA, *Consumer ADR in Europe*, Hart – C.H. Beck – Nomos, Oxford 2012.

[18] K. MANIA, 'Online dispute resolution: The future of justice' (2015) 1 *International Comparative Jurisprudence* 76, 83.

[19] For instance, Borlaw – Organismo di mediazione (Italy), Le service de médiation pour les télécommunications (Belgium) or CIAB – Centro de Informação, Mediação e Arbitragem de Consumo (Portugal).

[20] The mediator is 'any third person who is asked to conduct a mediation in an effective, impartial and competent way, regardless of the denomination or profession of that third person in the Member State concerned and of the way in which the third person has been appointed or requested to conduct the mediation'.

by the parties or suggested or ordered by a court or prescribed by the law of a Member State' (Article 3).

This very wide definition allows for the inclusion of almost all consumer ADR procedures, although greatly diverse, as long as they are not binding on the parties.

Consumer mediation has distinctive characteristics when compared with mediation in general, given the specificity of the disputes it aims to solve.[21] Consumer conflicts are essentially characterised by frequent inequality between the parties, usually a low value in dispute, and by the fact that, as a rule, there is no personal relationship between the parties and no need to maintain that relationship in the future.

It seems, however, that consumer mediation fits into the definition of Article 3 of Directive 2008/52/EC. In relation to the principles that complement the definition, some doubts may arise. Regarding the impartiality principle, one may argue that the obvious proximity to the consumer, bearing in mind that these entities are thought to ensure the effectiveness of the consumer rights, inevitably leads to a lack of impartiality on the part of the mediator.[22] However, we do not think that is the case. On the one hand, the effectiveness of consumer rights is ensured through the provision of means that are accessible and adequate for consumer claims and not by being biased towards consumers within those procedures. On the other hand, consumer ADR entities not only ensure the effectiveness of consumer rights, they also stimulate the functioning of the market, in that by boosting consumer trust, consumption rises. When the procedure begins, the role of the mediator is to listen to both parties on the subject matter of the mediation and allow them to determine – freely – the solution they consider most suitable for their dispute. Given that mediation is a voluntary procedure, a suggestion of bias would never be accepted by the other party, which would drop out of the process if they felt that the mediator was not being impartial. Partiality on the part of the mediator would thus lead to consumer mediation constantly failing.

Binding procedures may also be very diverse. On the one hand, they may be binding only for one of the parties, namely the trader, or for both. On the other hand, they may lead to a final decision at the end of a legal proceeding, as in arbitration, or a mere binding opinion, in which case the formality standards are lower.

---

[21] J. CAMPOS CARVALHO and J. MORAIS CARVALHO, 'Problemas Jurídicos da Arbitragem e da Mediação de Consumo' (2016) *Revista Electrónica de Direito* 1, 26.
[22] On the contrary, J. LUZAK, 'The ADR Directive: Designed to Fail? A Hole-Ridden Stairway to Consumer Justice' (2016) 24 *ERPL* 81, 95 points out the reasons why the consumer could fear the partiality of ADR procedures.

3.4. COSTS OF USE OF THE PLATFORM AND OF CONSUMER ADR PROCEDURES

The use of the platform is free for the parties (Article 5(1) of the Regulation), which does not necessarily mean that the consumer ADR procedures are also free. Article 9(7) of the Regulation determines that the ADR entity shall inform the parties 'of the costs of the dispute resolution procedure concerned'. This strengthens what arises from Article 7(1)(l) of Directive 2013/11/EU, which determines that ADR entities shall inform of 'the costs, if any, to be borne by the parties, including any rules on awarding costs at the end of the procedure'. Recital 41 of the Directive shows that the purpose is to have procedures that are close to free, noting that 'in the event that costs are applied, the ADR procedure should be accessible, attractive and inexpensive for consumers' and that 'costs should not exceed a nominal fee'.

Checking the list available within the platform, the conclusion is that in some cases costs are rather significant. Some ADR entities seem to apply the fees of commercial mediation or commercial arbitration, leading to fees that are greatly in excess of €1,000.[23]

In other cases, fees are applicable to the trader but not to the consumer,[24] or there are different costs that are higher for the trader. Is it lawful to charge different fees to both parties to the dispute? At first sight this practice could jeopardise the principle of equality – the paramount principle of due process – that should also guide ADR. In fact, the principle of equality is not one of the principles set forth in Directive 2013/11/EU, and there are reasons related to the nature of consumer disputes and the existing economic inequality between the parties that may justify the different treatment. Thus, the principle of equality does not seem to be violated in these cases; charging different fees to both parties is allowed as long as the fee charged to the consumer is low.

Several ADR entities do not charge fees to either party.[25] In these cases, the funding of ADR entities does not come from the parties, even in part. Two funding models can be found in this context: private funding, where consumer or trader associations or traders finance the system; and public funding, where the state provides, in a more or less centralised way, a free ADR system for both consumers and traders.

---

[23] See e.g. Sportello di conciliazione della Camera di commercio, industria, artigiano, agricoltura di Cosenza (Italy).

[24] For instance, Médiateur des communications électroniques (France), Allgemeine Verbraucherschlichtungsstelle des Zentrums für Schlichtung e. V. (Germany), ADR Group (UK) or NetNeutrals EU Ltd (Ireland).

[25] For instance, Centro Nacional de Informação e Arbitragem de Conflitos de Consumo (Portugal), Le service de médiation pour les télécommunications (Belgium) or Verein Internet Ombudsmann – nur für im Internet abgeschlossene Verträge (Austria).

This gives the parties an incentive to use the system, which can have a dual effect: on the one hand, it ensures disputes are solved that, due to their low value, would otherwise not come before an impartial and independent third party; on the other hand, it frees state courts from some disputes that can be better solved by other means, such as mediation.

3.5. COMMON RULES APPLICABLE TO THE ADR PROCEDURE WITHIN THE ODR PLATFORM

The procedure on the ODR platform begins with a claim submitted, in most cases, by the consumer.[26]

The complainant party submits the claim by filling in an electronic complaint form. The statement, in Article 8(1), that the complaint form on the ODR platform must be user-friendly and easily accessible, although very significant from a political point of view as regards consumer protection, has in fact a reduced normative content. The European Commission, which issued the proposal for the Regulation and was responsible for the creation of the ODR platform, has thus bound itself to create a form that is easily accessible and user-friendly.[27] We dare say that it would be odd not to aim at that, even without this self-imposed obligation. And what happens if it turns out that the form is actually not easily accessible and user friendly? Nothing. There are no consequences for the Commission failing to achieve this. If we add in the difficulty of knowing the exact meaning of such undefined concepts as 'easily accessible' and 'user-friendly', we come to the conclusion that we can but hope that the European Commission is set on creating an easy form. The definition of the level of simplicity must take into account the range of consumers that may access the platform, with very diverse technical and legal knowledge. It should not be assumed, for this purpose, that the consumers have average technical and legal knowledge, even though we are talking about disputes arising from contracts concluded online. It is in fact possible that many of these disputes arise precisely from the lack of specific knowledge, namely technical skills, at the time of concluding contract. Making it difficult for these consumers to access the platform would considerably reduce its effectiveness.

---

[26] As mentioned, dispute claims may also be filed by traders against consumers (see section 3.1).

[27] Article 2 of Implementing Regulation 2015/1051 complements Article 8(1) stating: 'The electronic complaint form to be submitted to the ODR platform shall be accessible to consumers and traders in all the official languages of the institutions of the Union. The complainant party shall be able to save a draft of the electronic complaint form on the ODR platform. The draft shall be accessible and editable by the complainant party prior to submission of the final fully completed electronic complaint form. The draft of the electronic complaint form that is not fully completed and submitted shall be automatically deleted from the ODR platform six months after its creation'.

Article 8 of the Regulation establishes the principles of adequacy and proportionality of the data provided by the consumer. This means data that are not useful for the handling of the case should not be required. It should be noted that it can be hard to define exactly what data will be necessary in the course of the procedure. For instance, an ID number is, as a rule, sufficient to ensure the consumer can be identified by the trader, but in some situations other elements such as a client number or a receipt number may be necessary for that purpose. To ensure the procedure is efficient, these data should ideally have been provided by the consumer when the claim is submitted.

The information that makes it possible to determine which ADR entity is competent to handle the procedure is especially important. Some entities may be competent for all kinds of disputes, but others are not. Besides entities with sectorial competence, others have their competence geographically defined. In these cases, it can be important to know the consumer's domicile or the place of fulfilment of the contract (to where the goods should be shipped or where the services should be provided).

The platform also allows for the sending of documents. This ensures the fulfilment of the due process principle, established in all Member States and also at the European level, as regards the right to present evidence. Especially in cases where the decision of the ADR entity is binding, it is of the utmost importance that the parties are allowed to present all the elements they deem important, allowing the third party to take them into consideration.

Once the claim has been filed, the ODR platform shall transmit it automatically to the respondent party[28] – in most cases the trader. The claim shall be transmitted 'in an easily understandable way and without delay' (Art. 9(3) of the Regulation). This rule is once again directed at the European Commission and uses undefined concepts, which raises the same problems identified above regarding the complaint form.

It is important to refer to the problem of the language(s) on the ODR platform. The claim is transmitted to the trader 'in one of the official languages of the institutions of the Union chosen by that party'. The trader may not have previously chosen one or more languages, as it may be the first time he accesses the ODR platform. This means that when the trader is notified of the claim submitted by the consumer he must choose the language or languages in which he wishes to interact with the platform. It may or may not be the language chosen by the consumer.

For that purpose, the ODR platform has an automatic translation function for all European Union official languages. According to recital 19 this electronic translation function should enable 'the parties and the ADR entity to have the information which is exchanged through the ODR platform and is necessary for

---

[28] A standard electronic message shall be sent by the ODR platform (Article 3 of Implementing Regulation 2015/1051).

the resolution of the dispute translated, where appropriate. That function should be capable of dealing with all necessary translations and should be supported by human intervention, if necessary'. This translation tool, if it is regularly complemented with new information and perfected, can could of great value for the European Union's activity, providing a database of words, expressions and sentences in several languages and their translation into all others.

The claim reaches the trader along with several elements listed in the subparagraphs of Article 9(3).

Firstly, the trader is informed that the parties have to agree on an ADR entity in order for the complaint to be transmitted to it, and that if no agreement is reached by the parties or no competent ADR entity is identified, the complaint will not be processed further. If the consumer refers to any ADR entity or entities which are competent to deal with the complaint in the electronic complaint form or if any entities are identified by the ODR platform on the basis of the information provided in that form,[29] that information is also transmitted to the trader, which makes his choice easier.[30] This rule needs to be adapted in cases where the trader is obliged, by national law, to use a specific ADR entity or to use the ADR entity chosen by the consumer, to resolve disputes with consumers. From a technical point of view the platform may not be prepared to handle these cases.[31]

Secondly, the platform sends out an invitation to the trader to state, within ten calendar days, whether he is committed or obliged to use a specific ADR entity to resolve disputes with consumers, and, if not, whether he is willing to use any of the competent ADR entities.[32]

Thirdly, the trader receives the name and contact details of the ODR contact point in the Member State where he is established or resident, as well as a brief description of the functions that are, in general terms, those of providing support and information to the parties.

Upon receipt by the respondent party, the ODR platform is programmed to communicate to the complainant party the same information it had previously sent to the respondent party (normally the trader), along with information about the ADR entity or entities chosen by the trader and 'an invitation to agree within 10 calendar days on an ADR entity' (Art. 9(4)(b)).[33]

---

[29] According to Article 4(1) of Implementing Regulation 2015/1051, 'the ODR platform shall display to the respondent party an indicative list of ADR entities, where no competent ADR entity is identified in the electronic complaint form, to facilitate the identification of the competent ADR entity'. This list shall be based on the geographical address of the parties and on the sector that the dispute relates to.

[30] According to Article 9(5), the listing of an ADR entity must be accompanied by a description of its characteristics.

[31] See section 4.2.

[32] Article 9(3) refers to the situation where it is the trader who submits the claim.

[33] Article 9(4)(c) refers to the situation where it is the trader who submits the claim, in which case the claim returns to him at this point to take a position on the choice made by the consumer.

The consumer must then choose one ADR entity from among those chosen by the trader. Once the consumer has made his choice, an agreement on the ADR entity has been reached and the ODR platform transmits the complaint to that ADR entity. It should be noted that, up until this moment, the ADR entity has no knowledge of the case nor that it might have to handle that dispute.

Next, the ADR entity shall inform the parties about whether it agrees or refuses to deal with the dispute. This information shall be given 'without delay', which in our opinion does not prevent the ADR entity from asking for additional information that is deemed important to make its decision. This decision is not a discretionary one. Besides the grounds of incompetence related to the subject, the territory or the value, there is an exhaustive list of grounds that the ADR entities on the platform may invoke to refuse to deal with a dispute. Whether these entities have adopted these grounds for refusal is stated on the platform and this information is transmitted to the parties when they choose the ADR entity, according to the selection procedure explained above. Article 5(4) of the Directive set out the following grounds for refusal: (a) the consumer did not attempt to contact the trader concerned to discuss his complaint and seek, as a first step, to resolve the matter directly with the trader; (b) the dispute is frivolous or vexatious; (c) the dispute is being or has previously been considered by another ADR entity or by a court; (d) the value of the claim falls below or above a pre-specified monetary threshold; (e) the consumer has not submitted the complaint to the ADR entity within a pre-specified time limit, which shall not be set at less than one year from the date upon which the consumer submitted the complaint to the trader; or (f) dealing with such a type of dispute would otherwise seriously impair the effective operation of the ADR entity.

The parties have a thirty-day period to reach an agreement concerning the ADR entity they want to choose. This time frame includes the ten days the trader has to reply after receiving the complaint, the ten days the consumer has to reply to the trader's proposal and ten additional days. These ten additional days show that the ten-day deadlines imposed on the trader and the consumer are not strict.

Where the parties fail to agree within thirty calendar days after submission of the complaint form or the ADR entity refuses to deal with the dispute, the complaint shall not be processed further. It should be noted that, where there is no agreement, the ADR entity does not even come to know of the existence of that dispute.

In these cases, the consumer is further informed of the possibility of contacting an ODR advisor for general information on other means of redress. The ODR advisors are hosted by the ODR contact points. According to Article 7(1) of the Regulation, each ODR contact point shall host at least two

ODR advisors. This is a way of avoiding the consumer becoming helpless and not knowing who to reach out to when faced with the trader's refusal.

The following figure outlines the various stages of the procedure.

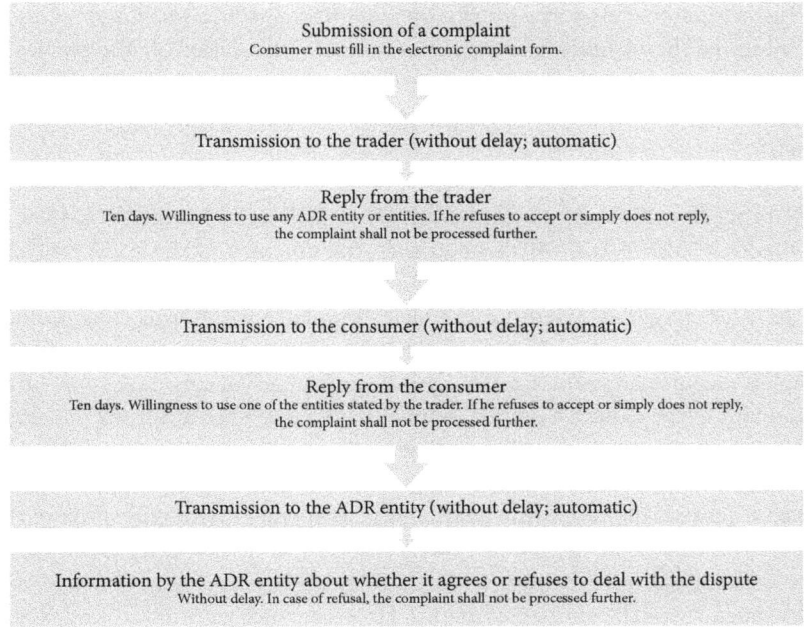

If the ADR entity agrees to deal with the dispute, it shall inform the parties of its procedural rules and costs.

From this moment onwards and up until the conclusion of the procedure it is not mandatory to use the platform.[34] According to Article 10(d), it is up to the ADR entity to decide whether or not to conduct the procedure through the ODR platform or through other means, namely those it usually uses for conducting the procedures outside the platform.

If the ADR entity chooses to conduct the procedure through the ODR platform it has a managing tool, which is cost-free for the parties. The procedure can be conducted entirely through the platform, or partly, the rest being conducted through the use of traditional means. It is, for instance, possible to do everything through the platform with the exception of an in-person hearing, if the person responsible for conducting the procedure deems it necessary. It should be noted, however, that the ADR entity shall not require the physical presence of

---

[34] S. PASSINHAS, 'Alterações Recentes no Âmbito da Resolução Alternativa de Litígios de Consumo' in A. PINTO MONTEIRO (ed.), *O Contrato – Na Gestão do Risco e na Garantia da Equidade*, Instituto Jurídico – Faculdade de Direito da Universidade de Coimbra, Coimbra 2015, p. 357, 376, stresses that 'more than ODR, the platform offers a single point of entry for complaints and their subsequent transmission to an ADR entity'.

the parties or their representatives, unless (i) its procedural rules provide for that possibility and (ii) the parties agree (Art. 10(b) of the Regulation). The efficiency of the procedure and preferences regarding how it is conducted prevail in part over the right to produce evidence.

The use of the platform allows for paperless dispute resolution with the advantage of the in-built functions of automatic translation. If the parties are interested in being actively engaged in the procedure, the platform allows for a quick and effective procedure, following the European Union's purpose in this domain.

Regardless of whether or not the platform is used as a managing tool, the ADR procedure must be concluded within the time frame set by Article 8(e) of Directive 2013/11/EU, which states that 'the outcome of the ADR procedure is made available within a period of 90 calendar days from the date on which the ADR entity has received the complete complaint file'.[35] This time period of ninety calendar days may be extended by the ADR entity 'in the case of highly complex disputes', but the parties shall be informed 'of any extension of that period and of the expected length of time that will be needed for the conclusion of the dispute'. It is thus established that, in most cases, the maximum length of the procedure will be ninety days. However, some flexibility is allowed. It would make no sense if the procedure could not be suspended were one of the parties, namely the consumer, to ask for a little more time to settle on some specific issue without the intervention of the ADR entity. It is true that the suspension will worsen the ADR entities' statistics. However, such entities do not exist for statistical purposes but rather to solve the problems of people.

Once the procedure is concluded, the ADR entity shall, even if it did not use the managing tool, update the information related to the case on the platform. According to Article 10(c) of the Regulation the ADR entity must transmit the date of conclusion and the result of the ADR procedure.

Regarding the result, Article 9(1)(c) of the Directive establishes that the parties must be notified of the outcome of the ADR procedure and be given 'a statement of the grounds on which the outcome is based'. The outcome varies according to the ADR procedure, ranging from an agreement or lack of agreement in non-binding procedures, to a binding decision in the case of arbitration.

The national and European rules on the different ADR procedures apply to the outcome of the procedure. Thus, the national rules transposing Directive 2008/52/EC shall, as a rule, apply to mediation procedures and the national laws on arbitration are applicable to arbitration, namely the rules on the arbitral award.

---

[35] The ADR entity shall transmit to the ODR platform the date of receipt of the complete complaint file, which starts the mentioned ninety calendar days (Article 5 of Implementing Regulation 2015/1051).

Having concluded the analysis of the ODR platform and especially of the dispute resolution procedure, it is now time to perform a critical analysis of the main questions the ODR platform raises.

4. CRITICAL ANALYSIS OF THE LEGAL REGIME AND ITS IMPLEMENTATION

4.1. CONSUMER INFORMATION

The ODR platform relies on the awareness of its existence by the citizens of the Member States to fulfil its purpose as a fundamental instrument to ensure the enforcement of European contract law, and especially of the digital single market. If the consumer is not aware of the existence of the ODR platform he will never be able to use it to solve a dispute. Likewise, if the trader does not have information on ADR and ODR and about the possibility of solving disputes in an efficient manner through the platform, he will be suspicious upon receiving the claim and act defensively, which will hinder the procedure's chances of success.

Besides the general information on the platform provided by the European Commission[36] and the competent authorities in the Member States,[37] the Regulation obliges the trader to supply information.[38]

Article 14(1) of the Regulation sets forth that 'traders established within the Union engaging in online sales or service contracts, and online marketplaces established within the Union, shall provide on their websites an electronic link to the ODR platform' and that 'that link shall be easily accessible for consumers'. Where on the trader's website should the electronic link be placed? The answer to this question alone may determine the success or failure of the platform.

---

[36] Article 5(3) establishes that: 'The Commission shall make the ODR platform accessible, as appropriate, through its websites which provide information to citizens and businesses in the Union and, in particular, through the "Your Europe portal" established in accordance with Decision 2004/387/EC'.

[37] The ODR contact points are regulated in article 7 of the Regulation. Article 14(5) sets forth that Member States shall ensure that ADR entities, the centres of the European Consumer Centres Network, the competent authorities defined in Article 18(1) of Directive 2013/11/EU, and, where appropriate, the bodies designated in accordance with Article 14(2) of Directive 2013/11/EU provide an electronic link to the ODR platform'. Also Article 14(6) establishes that: 'Member States shall encourage consumer associations and business associations to provide an electronic link to the ODR platform'.

[38] There is no doubt that 'the information obligation may not be sufficient incentive for traders to participate in an ODR process' (P. CORTÉS and A.R. LODDER, 'Consumer Dispute Resolution Goes Online: Reflections on the Evolution of European Law for Out-of-Court Redress' (2014) 1 *Maastricht Journal of European and Comparative Law* 14, 30), but it certainly has a huge persuasive effect on the trader. I. BENÖHR, 'Alternative Dispute Resolution for Consumers in the European Union' in C. HODGES, I. BENÖHR and N. CREUTZFELDT-BANDA (eds.), *Consumer ADR in Europe*, Hart – C.H. Beck – Nomos, Oxford 2012, p. 1, 22 highlights the importance of this information.

On the basis of research conducted on company websites at the end of May 2016, it can be concluded that the information is not 'easily accessible'. In order to reach the information on the ODR platform on the website www.vertbaudet.com, it was necessary to open a scarcely visible link at the bottom of the page named 'Terms and Conditions'. The vast majority of consumers – not to say all – will not realise that the platform exists. On Amazon, it was even more difficult to find the information. We could only find it by searching the website for the expression 'dispute resolution'.[39] It is possibly also available through one or a chain of links accessible from the homepage but we were unable to find it. On the websites of Uber and Airbnb we could not find the information at all. It should be noted that, whether they qualify as service providers or mere intermediaries, online marketplaces are also obliged to provide the information on the ODR platform.

It is thus possible to conclude that even when the information is provided by the trader on its website it is just a formality. From a material point of view, the reality is exactly the same as if the information were not provided. We would argue that the present way in which the information is provided would not comply with the duty imposed by the Regulation even if the way in which information is provided were not regulated. The non-compliance is much clearer if we bear in mind that the Regulation states that the information must be 'easily accessible for consumers'.

The information is only easily accessible if it stands out when the consumer accesses the website or at least when a contract is concluded. The consumer should not have to actively do anything in order to get that information. It should for instance not be necessary for him to already be party to a dispute and be searching for ways to resolve it.

The best solution to this problem – which, we stress, is crucial for the success of the platform – would be to create a European logo or a logo for the platform that would have to be visible on the trader's website at all times, regardless of the type of device that is being used.

Besides representing a fairly objective standard that does not rely on undefined concepts such as ease of access, displaying the logo and it consequently being seen by the consumer when he first accesses the website effectively ensures he will know of the existence of the platform.

The logo is essential to ensure that the platform has a recognisable image, making it easily identifiable to users. Written information, without the use of images, goes unnoticed in today's context. If one examines the websites mentioned above it becomes clear that all commercial information related to the company's goods and services is combined with images or videos. The omission

---

[39] The information could, at the end of May 2016, be found here: <https://www.amazon.co.uk/gp/help/customer/display.html?ref=help_search_1-1?ie=UTF8&nodeId=1040616&qid=1464180508&sr=1-1>.

of these visual elements renders the message much less interesting and therefore less likely to be noticed by consumers.

This measure would have to be taken on a European level, because the logo cannot be created unilaterally by one or several Member States. The individual Member States could ensure that the information is accessible immediately upon opening the website, as it is a measure that fits into Article 18 of the Regulation, which mentions taking 'all measures necessary' to ensure the implementation of the Regulation. However, from a practical point of view, bearing in mind the internationalisation of a large number of traders, the imposition of different ways of providing information could cause problems.

4.2. MANDATORY ADR

Despite the reference in some rules of the Regulation to the possibility of the trader being committed to, or obliged to use, a specific ADR entity to resolve disputes with consumers, the platform is clearly not set up for these types of cases.

As we have seen, Article 9(3) requires the trader to state whether or not he 'commits to, or is obliged to use, a specific ADR entity to resolve disputes with consumers'.[40] However, if the trader is committed or obliged to use a specific ADR entity and omits that information, the case cannot be processed further on the platform. There may be a sanction imposed by the Member State, according to Article 18 of the Regulation, but the platform does not allow for the case to go on. For example, if the trader is obliged to use arbitration, either by commitment or by legal imposition, the arbitral procedure will not even be initiated on the platform. It is true that the consumer may initiate the procedure outside the platform, but a large part of the platform's purpose is still lost. In addition, where the trader does not reply to proceedings initiated via the platform,[41] in many cases the consumer will not know that the trader is bound by the procedure, which may lead him to give up the claim. The case ends up not being resolved because of the platform's ineffectiveness.

---

[40] Article 14(1) stresses this duty by establishing that 'traders established within the Union engaging in online sales or service contracts, which are committed or obliged to use one or more ADR entities to resolve disputes with consumers, shall inform consumers about the existence of the ODR platform and the possibility of using the ODR platform for resolving their disputes'. It also states that 'they shall provide an electronic link to the ODR platform on their websites and, if the offer is made by e-mail, in that e-mail. The information shall also be provided, where applicable, in the general terms and conditions applicable to online sales and service contracts'.

[41] The option of not replying may not be a clever strategy bearing in mind the client's satisfaction. See S. BLAKE, J. BROWNE and S. SIME, *A Practical Approach to Alternative Dispute Resolution*, 3rd ed., Oxford University Press, Oxford 2014, p. 15.

The solution to this problem could be a change in procedure, whereby the ADR entity would screen all cases after the claim has been presented and before transmitting it to the trader and giving him the opportunity to state whether he agrees or refuses to take part in the procedure. In cases where his participation in the procedure is mandatory, the question would not even be raised. The claim would be transmitted to the trader merely to inform him that the procedure has been initiated and what the subsequent stages will be.

4.3. PERSUASIVE EFFECT OF THE ADR ENTITY'S INTERVENTION

Closely connected to the previous question, we think that the main obstacle to the success of the platform is the fact that the ADR entity is not able to intervene in the early stages of the procedure.

The way the platform is designed allows traders to choose to ignore claims they receive, regardless of whether or not they are obliged to use ADR. If the trader chooses to ignore the claim, he is both spared the costs of analysing the claim and responding to the consumer, and, because the claim is dropped, he no longer has to worry about it. Consumers may perhaps try other means as an alternative, but likely only a (very) few will do so, as the discouragement associated with the lack of response will be significant.

If the ADR entity were able to intervene early on, right after the submission of the claim, it could play a significant role in persuading the trader to take part in the procedure. Merely providing information on how ADR works can be decisive in many cases. A reminder of the possibility to reply to the invitation stating whether or not he is willing to make use of ADR would certainly drive more traders to take part.

If we do not wish the platform to have no cases to handle, it is vital that we trust the ADR entities (which, by law, are – or at least should be – independent and impartial) and their ability to be a key player in promoting contact between the parties within the system. This necessarily includes their importance at the stage where the parties are still deciding whether or not they wish to make use of ADR.

We wish to emphasise the importance of the ODR platform as an instrument to render European contract law more effective, in particular in the context of the digital single market, but fear that, without some small but decisive adjustments that can be made through secondary regulation, its usefulness will be rather limited.